The Economics of the Sulphur Industry

Between the 1950's and 1970's, the sulphur industry continued to grow despite occasional shortages and excesses. In this study originally published in 1970, Hazleton focuses on the Frasch sulphur industry to explore issues such as competing sources of sulphur, the possibilities of sulphur being obtained as a result of pollution-abating policies and the conditions under which future supplies are likely to become available. This title will be of interest to students of Environmental Studies.

The Economics of the Sulphur Industry

The Economics of the
Sulphur Industry

Jared E. Hazleton

First published in 1976
by Resources for the Future, Inc.

This edition first published in 2016 by Routledge
2 Park Square, Milton Park, Abingdon, Oxon, OX14 4RN
and by Routledge
711 Third Avenue, New York, NY 10017

Routledge is an imprint of the Taylor & Francis Group, an informa business

© 1976 Resources for the Future, Inc.

Publisher's Note
The publisher has gone to great lengths to ensure the quality of this reprint but
points out that some imperfections in the original copies may be apparent.

Disclaimer
The publisher has made every effort to trace copyright holders and welcomes
correspondence from those they have been unable to contact.

A Library of Congress record exists under LC control number: 76075181

ISBN 13: 978-1-138-95576-9 (hbk)
ISBN 13: 978-1-315-66604-4 (ebk)
ISBN 13: 978-1-138-95577-6 (pbk)

The Economics of the Sulphur Industry

The Economics of The Sulphur Industry

The Economics of the Sulphur Industry

Jared E. Hazleton

Published by Resources for the Future, Inc.
Distributed by The Johns Hopkins Press, Baltimore and London

Resources for the Future is a nonprofit corporation for research and
education in the development, conservation, and use of natural resources
and the improvement of the quality of the environment.
It was established in 1952 with the cooperation of the Ford Foundation.
Part of the work of Resources for the Future is carried out by
its resident staff; part is supported by grants to universities and
other nonprofit organizations. Unless otherwise stated, interpretations
and conclusions in RFF publications are those of the authors; the
organization takes responsibility for the selection of significant subjects
for study, the competence of the researchers, and their freedom of inquiry.

This book is one of RFF's resources appraisal studies, which are
directed by Hans H. Landsberg. Jared E. Hazleton is a lecturer
in economics at the University of Texas at Austin.
The charts were drawn by Clare and Frank Ford.

RFF staff editors: Henry Jarrett, Vera W. Dodds, Nora E. Roots, Tadd Fisher.

Foreword

Unless the next five years produce some unexpected changes, the Paley Commission's 1952 estimate that U.S. sulphur demand in 1975 would be twice that of 1950 will turn out to have been an accurate one. The reason is not hard to find: the use of sulphur has in the past so permeated the economy that the growth of the economy itself has pretty much set the pace of sulphur consumption. And in the past two decades both have grown at about three per cent per year. But, as Jared Hazleton shows, this smooth parallelism has been strained at various times during the twentieth century, with periods of sulphur shortage and glut following one another.

In this study—the revised version of a Ph.D. thesis that was supported by a grant under RFF's Fellowship Program—the author traces the characteristics of the Frasch sulphur industry, which has constituted the largest source of sulphur during this century. For his analysis of the output and price policies followed by the handful of producers that make up the industry, he has drawn upon interviews with industry representatives as well as upon the limited literature and the none-too-satisfactory published statistics.

This may be a particularly good time for such a study. One could make a reasonable guess that the Frasch sulphur industry has reached a turning point. Competing sources of sulphur are continuing to grow and are reducing the dominant role of the Frasch segment in the market. At the same time, the Frasch sulphur companies themselves have begun to diversify into other fields. Finally, we may be on the verge of a massive new source of supply—sulphur or sulphuric acid obtained as a result of pollution-abating practices only now coming into view. What such a development would spell for the conventional producers nobody has yet worked out; but that it would open a wholly new chapter in the life of an ancient industry is certain. In any event, Mr. Hazleton's book is of interest and value as a case study of industrial organization, an analysis of industrial behavior in the presence of great market power, and a contribution to an appraisal of the conditions under which future supplies are likely to become available from both tried and untried sources.

Hans H. Landsberg
Director of Resource Appraisals
Resources for the Future, Inc.

Acknowledgments

This study was first undertaken as a doctoral dissertation prepared for Rice University, under a Natural Resources Fellowship awarded by Resources for the Future. Subsequently, it has been extensively revised, updated, and the chapter on the industry and its resource has been added.

I am indebted to Professor Edward S. Mason of Harvard University for first suggesting to me that the sulphur industry would make an interesting topic for study. Professor Henry Steele, as director of my thesis committee, contributed freely of his time and advice. Professors Gaston Rimlinger and Dwight Brothers read and commented upon various drafts of the manuscript while it was at the dissertation stage.

I owe a special debt of gratitude to Hans H. Landsberg of Resources for the Future, who provided guidance and encouragement in the task of converting the dissertation to its present form. I also wish to thank Professor Lee E. Preston, Associate Dean of the School of Business Administration, University of California at Berkeley, who commented on an earlier draft.

As anyone who has undertaken an industry study realizes, this study could not have been made without the assistance and cooperation of a number of individuals in the sulphur industry, trade associations, and government, who gave freely of their time in providing me with information on the industry. They, of course, bear no responsibility for the conclusions of the study.

My final debt of gratitude is to my wife and family—they also serve who only stand and wait and wait and wait—to whom this book is dedicated.

Jared E. Hazleton

Table of Contents

List of Tables

List of Illustrations

The Economics of the Sulphur Industry

Chapter 1

Introduction

Sulphur is one of the most plentiful of the elements, but twice within the postwar era it has been in tight supply both within the United States and abroad. Sulphur is also one of the most important industrial raw materials, yet it is one of our least known mineral resources.

This study is an economic analysis of the sulphur industry. Its goal is to describe, interpret, and evaluate from the standpoint of public welfare the market structure, behavior, and performance of the domestic sulphur industry. At the same time, economic analysis of the industry should reveal much about the nature of the sulphur resource. In this regard, the study seeks to answer two questions. Does the past pattern of periodic shortage and abundance indicate recurring periods of scarcity in the future? Can the future sulphur requirements of the world be met without significant increases in unit costs?

Description of the Industry

Sulphur is widely distributed in nature, both alone and in combination with other minerals. Although 0.06 per cent of the weight of the earth's crust is believed to consist of sulphur, only an exceedingly small portion of the sulphur occurs in sufficiently concentrated amounts to justify mining. When considered as an industrial raw material, sulphur may be classified into two types. Where not in molecular combination with any other element, sulphur is called "elemental" and, in its natural state, is often termed "brimstone." Sulphur is also classified as elemental where, prior to its use as a raw material, it is separated in pure form from compounds. Sulphur is classified as "nonelemental" where, in molecular combination with some other element, it is used as a raw material without prior separation.

Elemental sulphur is supplied from several sources. Frasch sulphur, which derives its name from the developer of the process by which it is mined, is produced from brimstone deposits found in the salt domes of the United States and Mexican Gulf Coasts. The Frasch process is also used to produce elemental sulphur from non-salt dome deposits in West Texas. In addition, a small quantity of elemental sulphur is mined by conventional methods from sedimentary deposits and surface deposits, and separated from the metal in metallic sulphides. Recovered sulphur is produced from the hydrogen sulphide contained in sour natural or refinery gases.

Nonelemental sulphur is supplied principally as by-product acid from copper and zinc smelters, from gypsum and anhydrite deposits, and from pyrites (metal sulphides), which are burned to produce sulphur dioxide and by-product metallic oxides. A small amount of nonelemental sulphur is produced in the form of hydrogen sulphide by oil refineries and in the form of sulphur dioxide by smelters.

As shown in table 1, domestic production of sulphur in all forms totaled over 9.1 million tons in 1967.[1] Frasch-produced sulphur accounted for 7 million tons, or 75 per cent of total domestic production. Recovered sulphur output of nearly 1.3 million tons accounted for 14 per cent of domestic sulphur production, and various other sources accounted for the remaining 11 per cent. The total value of domestic sulphur shipments in 1967 amounted to about $292.8 million.

Table 1.   U.S. Sulphur Production, All Forms, 1967

Source	Net sulphur content
	long tons
Elemental sulphur:	
Frasch process mines	7,014,164
Other mines	284
Recovered	1,267,955
Total elemental sulphur	8,282,403
Nonelemental sulphur:	
Pyrites (including coal brasses)	355,033
By-product	498,675
Total nonelemental sulphur	853,708
Total U.S. sulphur production	9,136,111

SOURCE: U.S. Department of the Interior, Bureau of Mines, *Minerals Yearbook*, 1967.

1. All statistics dealing with physical quantities in the industry will be given in long tons (2,240 pounds) unless otherwise noted.

The Approach

Industry studies by economists have generally been directed toward the analysis of imperfections in the competitive economy, primarily toward the analysis of oligopolistic markets where a few sellers possess varying degrees of market power. Market power exists where at least one firm is of sufficient size relative to the market to influence by its own actions the level of market price and output. The existence of market power results in a situation in which the equilibrium price and output in the industry are indeterminate. Thus, the task of an economic study seeking to analyze an industry of few sellers is to explain why the pattern of price and output behavior observed in the market emerged from the oligopoly being studied, and to provide a full critique of the impact of this behavior on the public welfare.

The sulphur industry is a typical oligopoly in which a few firms have consistently controlled a large share of the market. This study of the sulphur industry is organized along the lines of a number of similar studies of oligopolistic industries. It focuses on the industry's market structure, behavior, and performance.

This approach postulates that definable factors of market structure interact to produce a pattern of market behavior on the part of firms in an industry. This pattern of market behavior in turn provides a basis for evaluating the performance of the industry. The purpose of appraising the performance of the industry is to establish the extent to which firms have used their market power to achieve noncompetitive results. The explanation of market behavior in terms of the market structure of the industry is designed not only to describe how the pattern of market behavior emerged from the industry, but also to identify alternative market structures, obtainable through public policy, which might yield a better pattern of performance for the industry.

The market structure-behavior-performance approach attempts to solve the problem of oligopolistic indeterminacy by introducing additional factors of market structure into the analysis. Thus, structure is defined to include all factors taken into account by the firm in determining its business policies and practices. While the number and relative sizes of the sellers and buyers are an important element in this regard, they represent only one of several aspects of market structure.

Other factors that can be identified as exerting an influence on the firm's behavior include: (1) the geographical distribution of supply and demand; (2) the nature of demand, including the elasticity of demand with respect to both prices and income in the short run as well as in the long run; (3) the organization of the firm, in particular, the marketing and distribution channels used; (4) the nature of costs, including the proportion of fixed costs in the short run, and the ease of exit from the industry; (5) the conditions affecting entry into the industry; (6) the technological environment, i.e., the limitations imposed on the firm's actions by the current state of technology; and (7) the legal setting, including the impact of the patent system, taxation, and other governmental policies and regulations which affect the firm's behavior.[2] While this list is not exhaustive, it summarizes the more important features that lend themselves to empirical analysis.

The oligopolist, operating within an environment defined in terms of these factors of market structure, adopts certain policies in the market where it sells its product. Market behavior is formed by these policies of business firms in determining such factors as prices, outputs, product characteristics, selling expenses, and research expenditures. The importance of market behavior lies in its function as the link between an industry's structure and its performance.

In an oligopolistic industry, market behavior reflects the firm's policies not only toward its product market but also toward the actions taken by its rivals in that market. It is this interreaction of firms in the market that gives oligopolistic industries their unique indeterminateness and makes them an attractive subject for economic analysis. One important set of market behavioral patterns centers on determining oligopolistic prices, i.e., the ways in which firms set their prices and change them in response to others. A second set of policies concerns the product of the oligopolist. Product policies include product differentiation, the level of selling costs, and product quality. The final set includes policies that seek to change the factors of market structure. Coercive practices, for example, seek to drive out weaker rivals, or to reduce the threat of entry. Fear of antitrust litigation may have the opposite effect and result in policies aimed at maintaining the market share of the smaller firms in the industry.

The third element of industry analysis, market performance, is closely related to market behavior.[3] Behavior provides a basis for an evaluation of industry performance in terms of certain economic criteria. These criteria include: (1) cost-price relationships: the extent to which reduc-

---

2. These factors are summarized from those mentioned by Merton J. Peck, *Competition in the Aluminum Industry* (Cambridge: Harvard University Press, 1961), p. 2; Charles H. Phillips, Jr., *Competition in the Synthetic Rubber Industry* (Chapel Hill: The University of North Carolina Press, 1961), p. 6; and James W. McKie, *Tin Cans and Tin Plate* (Cambridge: Harvard University Press, 1959), p. 5.

3. The term "behavior" is often used to include both the conduct of the firms in the industry and the performance of the industry. It is desirable, however, to distinguish between these two elements because conduct is subject to objective analysis, while performance is substantially a matter of subjective evaluation.

tions in cost are passed on promptly in the form of price reductions; (2) capacity-output relationships: the efficiency of resource use as indicated by the scale and utilization of facilities and the location of production; (3) progressiveness: the extent to which the firms in the industry actively and effectively engage in product and process innovation; and (4) the level of profits: the extent to which profits of firms in the industry (rate of return after taxes on invested capital) exceed or fall short of profits earned by firms in industries exhibiting similar trends in such factors as sales, costs, and innovations.[4]

The purpose of appraising industry performance is to indicate the impact of market power on the public welfare. This impact is usually described in terms of the effectiveness or workability of competition. Economists are in general agreement that perfect competition constitutes neither a normative ideal nor a satisfactory basis for appraising actual market conditions. Therefore, they have sought to derive a "workability criterion," a set of conditions that would be both necessary and sufficient to ensure that market power is used in the public interest.

Unfortunately, it is easier to recognize the need for such normative standards than to provide broadly applicable criteria. At best, the appraisal of the workability of competition in an industry remains a "subjective judgment by a given economist concerning the extent to which he thinks that absence of one or another of the conditions of perfect competition will not prove *unduly* harmful to economic welfare."[5]

An appraisal of workability may lead to one of three conclusions. First, it may be determined that market power is present but has not been used to secure noncompetitive results, and the industry is judged workably competitive. Second, market power may be shown to have produced imperfectly competitive performance which is considered nonetheless workably competitive in light of possible remedial actions available through public policy. Third, market power may be shown to have produced noncompetitive performance that is considered unsatisfactory, and that can be improved through use of public policy. In the latter instance, the industry is considered nonworkably competitive. The key elements in any evaluation of the workability of competition in an industry are, therefore, the degree to which actual performance deviates from desired performance, and the possibility of remedial public action.

4. This list is taken from Edward S. Mason, "The Current Status of the Monopoly Problem," *Harvard Law Review* (June 1949), pp. 1281–82; and Carl Kaysen, *United States v. United Shoe Machinery Corporation* (Cambridge: Harvard University Press, 1956), p. 17.

5. H. H. Liebhafsky, *The Nature of Price Theory* (Homewood, Illinois: The Dorsey Press, Inc., 1963), p. 22.

Application to a Resource Industry

The considerations involved in an industry study described above are particularly suited to an analysis of a resource-based industry such as sulphur. Like other minerals, sulphur is characterized by fixed location, exhaustion of reserves, changing patterns of supply, competition between primary and by-product production, and continuously changing technology. Like other minerals, sulphur has periodically been the subject of public concern because of high prices and inadequate supply.

Harold J. Barnett and Chandler Morse have considered extensively the problem of increasing natural resource scarcity and diminishing returns as reflected in the trend of unit costs of extractive products. They conclude that to a significant degree, the prospect of increasing resource scarcity often generates its own solution in the form of sociotechnical change which results in increasing as opposed to diminishing returns.[6] There is a marked similarity between the argument as advanced by Barnett and Morse and the earlier theory of "creative destruction" developed by Joseph Schumpeter.[7] Both emphasize the dynamic nature of the capitalist economy. Both stress that economic change is embodied in sociological and technological change.

But what is the process through which sociotechnical change occurs? What determines the speed with which advances in technology act to alleviate problems of resource scarcity? How automatic is the process? The fundamental element in the process is, of course, the market, which both signals the need for change and provides the reward which motivates change. New sources of supply, new products and processes, and new forms of organization affect the market structure of an industry. At the same time, they are affected by changes in market structure and by changes in the behavior and performance of the industry.

Thus, a primary goal of this study will be to define the relationship between the sulphur industry and the sulphur resource. In particular, the study will investigate the role of market structure in fostering or hindering extension of the sulphur resource base. Extensions of the resource base will be analyzed to determine if they originated within or without the industry. Their impact on market behavior will be examined. Industry performance will be evaluated not only from the standpoint of

6. Harold J. Barnett and Chandler Morse, *Scarcity and Growth* (Baltimore: The Johns Hopkins Press for Resources for the Future, Inc., 1963). Barnett and Morse first emphasize (p. 7), "Resources can only be defined in terms of known technology." They then conclude (p. 10), "the increasing scarcity of particular resources fosters discovery or development of alternative resources, not only equal in economic quality but often superior to those replaced."

7. Joseph A. Schumpeter, *Capitalism, Socialism, and Democracy* (New York: Harper and Row Publishers, Inc., 1942).

the competitive norm but also in light of the industry's effectiveness in exploiting the sulphur resource.

Plan of This Study

This study is divided into three parts along lines similar to the market structure-behavior-performance approach outlined above. Part I describes the market structure of the sulphur industry. Chapters 2 and 3 deal with the supply aspects and chapter 4 with the demand aspects. Part II analyzes market behavior in the industry. Chapter 5 outlines the behavior of output and prices, while chapter 6 provides a chronological explanation of market behavior from the industry's inception to the present. In Part III, chapter 7 looks at performance from the standpoint of the competitive norm. Chapter 8 appraises the industry's performance with regard to the sulphur resource and presents an evaluation of the outlook for sulphur as a resource. In the final chapter, the major findings of the study are summarized, and the workability of competition in the industry is assessed.

Part I

Market Structure of the Sulphur Industry

Chapter 2

Sources of Supply of Sulphur

Although this study is primarily concerned with the Frasch segment of the domestic sulphur industry, any adequate analysis of the industry must also consider the competitive potential of other sources of supply. This chapter traces the development of the Frasch sulphur industry in both the United States and Mexico, and then describes the recovered sulphur industry which has become an increasingly important source since the Second World War. Other sources of supply which contribute only a small fraction of total sulphur supply are not discussed.

Frasch Sulphur

While sulphur has been known and used since antiquity, it was not until the last decades of the eighteenth century that advances in chemical technology centering around sulphuric acid led to increased demand for sulphur for industrial use. During the early years of the nineteenth century, the world's sulphur needs were supplied by the large volcanic deposits of elemental sulphur in Sicily.[1] The disadvantages of relying on a single source became obvious when the King of Naples granted a monopoly to a French firm for the sale of Sicilian sulphur in 1838, and the company promptly increased the price of sulphur from $25 to $75 per ton, the maximum permitted under its contract. Pressure from foreign purchasers, particularly Great Britain, forced a rescission of the contract in 1840, but by that time an interest in new sources of supply had been aroused. By 1860, Sicily still supplied the bulk of the world's nonacid sulphur needs, but acid manufacturers outside the United States had turned to obtaining sulphur from the roasting of iron pyrites.

1. For a complete description of the geological nature of the Sicilian sulphur deposits, see Walter F. Hunt, "The Origin of the Sulphur Deposits of Sicily," *Economic Geology* (1915), pp. 543–79. The most comprehensive description of the industry in Sicily is found in William Haynes, *The Stone That Burns* (New York: D. Van Nostrand Company, Inc., 1942).

At the turn of the century, American acid manufacturers also switched to iron pyrites. Interestingly enough, this action corresponded with the beginning of the Frasch sulphur industry in the United States.

Description of the Frasch process[2]

The Frasch process generally is applicable to a specific type of sulphur deposit associated with geological formations known as salt domes. In the United States, salt domes occur within a wide belt reaching from Alabama to Texas and extending offshore into the Gulf of Mexico and 75 to 300 miles inland. Outside the United States, salt domes have been discovered in Germany, the Netherlands, Romania, Iran, Russia, and Mexico. Only the U.S. and Mexican deposits have proven amenable to Frasch sulphur mining.

Basically, salt dome formations are pillars of salt extending to great depths and forming a dome-shaped structure in the overlying sediments. The tops of the pillars contain a characteristic series of minerals known collectively as the cap. Cap rock formations, consisting of limestone, gypsum, and anhydrite, differ materially in size, depth, thickness, and configuration, as well as in the relative proportions of the principal and accessory mineral constituents. Within the cap rock, sulphur occurs as well-developed rhombic crystals in the fissures, cracks, and seams of the porous limestone and also in the semicrystalline or massive state as a filling in the openings of formations. In a productive dome, the sulphur-bearing limestone is sealed in by a layer of barren, nonsulphur-bearing limestone above and a layer of equally barren anhydrite or gypsum below. The thickness of the sulphur formation varies widely, though the average is about 100 feet. The average sulphur content in a deposit ranges from 20 to 40 per cent.

The Frasch process for mining sulphur is based on the fact that crystalline sulphur melts at about 116° C. In mining, wells are drilled into the sulphur formation using rotary rigs similar to those employed in the

2. This description of salt dome sulphur deposits and the Frasch process is based on that given by Paul M. Ambrose, "Sulphur and Pyrites," in *Mineral Facts and Problems, 1965 Edition, Bulletin 630*, Bureau of Mines, U.S. Department of the Interior, Washington, D.C., 1965. In addition, for descriptions of salt dome sulphur deposits see: Marcus A. Hanna and Albert G. Wolf, "Texas and Louisiana Salt Dome Cap Rock Minerals," and George Sawtelle, "Salt Dome Statistics," in *Geology of Salt Dome Oil Fields*, ed. Raymond C. Moore (Tulsa, Oklahoma: The American Association of Petroleum Geologists, 1926), pp. 119–32 and 109–18; and W. T. Lundy, "Sulphur and Pyrites," in *Industrial Minerals and Rocks* (2nd ed.; New York: American Institute of Mining and Metallurgical Engineers, 1949). The most complete account of the development of the Frasch process appears in William Haynes, *The Stone That Burns*, pp. 24–45. A good description of the technical aspects of mining sulphur by the Frasch process is given by Will H. Shearon, Jr., and J. H. Pollard, "Modern Sulphur Mining," in *Modern Chemical Processes*, ed. William J. Murphy (vol. 2; New York: Reinhold Publishing Corporation, 1952), pp. 219–29.

petroleum industry. The wells are equipped with a nest of four concentric pipes. A typical well will have an outermost pipe, eight or ten inches in diameter, which extends to the top of the cap rock and lines the well to prevent the sides from caving in. Inside this outer pipe is a six-inch pipe which extends through the sulphur-bearing stratum and rests in the upper portion of the underlying anhydrite or gypsum layer. A three-inch pipe, placed inside the six-inch pipe, reaches nearly to the bottom of the sulphur-bearing rock, but rests on a collar that is set within the six-inch pipe and that seals the annular space between the two. Finally, a one-inch air pipe, inside the three-inch pipe, extends to an area slightly above the collar. The six-inch pipe is perforated at two levels, separated by the collar; the upper set of holes permits the escape of hot water and the lower set the entrance of molten sulphur (see figure 1).

In steaming a well, water is treated to remove scale-forming salts and corrosive substances; heated to 320°–340° F.; pumped, under pressure

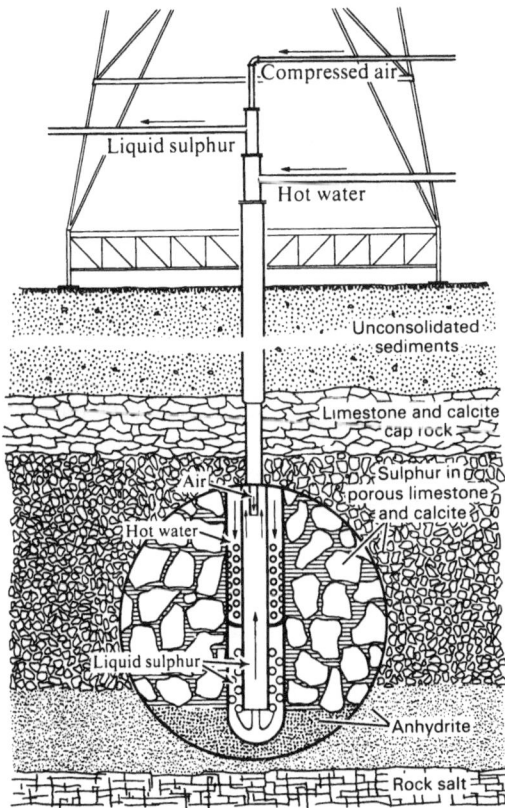

Figure 1. Underground piping arrangement for Frasch mining.

of 100–250 pounds per square inch, down the annular space between the six-inch and three-inch pipe; and discharged into the porous limestone formation through the upper set of perforations in the six-inch pipe. The entire region through which the water circulates is raised to a temperature above the melting point of sulphur. The heated zone formed by the incoming water takes the shape of an inverted cone, whose outer limits remain below the fusion point of sulphur and whose apex is the bottom of the hole. The molten sulphur, being heavier than water, runs within the dome. The amount of water required to recover a ton of sul-Dome pressure forces the sulphur part-way up the three-inch pipe. Compressed air, injected at a pressure of about 500 pounds per square inch through the central one-inch pipe into the molten sulphur, changes the apparent specific gravity of the liquid sulphur and raises it to the surface. Liquid sulphur reaching the surface is discharged into steam-heated tanks. After being metered, the sulphur is pumped through steam-heated lines into vats, where it is stored in either dry or liquid form.

Since each well is able to remove sulphur from only a limited area, the Frasch process requires the constant drilling of new wells. In addition, bleeder wells must be drilled to remove the large volumes of cool water that accumulate within the formation; this makes room for the hot water being pumped into the dome and prevents a buildup of pressure to the bottom of the cone, and forms a pool around the foot of the well. phur varies widely between domes.

The removal of large quantities of sulphur from the cap rock usually affects production in one of two ways. If cavities remain after the sulphur is removed, there is an increase in the water-to-sulphur ratio and, therefore, in the costs of production. To reduce hot-water requirements, the cavities are filled by mudding. If sulphur removal is followed by subsidence, the cavities are often filled and there is no need for mudding, but subsidence often distorts or breaks the strings of pipe in the well and damages surface installations in the immediate area. The defense against subsidence is to drill and equip wells only as needed and to construct power plants and other permanent structures at points beyond the limits of any possible surface subsidence area.

Development of the U.S. Frasch sulphur industry

The birth of the Frasch sulphur industry in the United States can be dated from 1868, when an extensive sulphur deposit was discovered by the Louisiana Petroleum and Coal Oil Company in Calcasieu Parish, Louisiana, about 30 miles inland from the Gulf of Mexico.[3] The firm,

---

3. This section draws extensively on the descriptions of the U.S. Frasch sulphur industry given by William Haynes in *The Stone That Burns* and in *Brimstone: The Stone That Burns* (Princeton: D. Van Nostrand Company, Inc., 1959).

while prospecting for oil, had discovered a bed of pure sulphur, 108 feet thick, at a depth of about 500 feet. Between 1870 and 1890, several unsuccessful attempts were made to mine this deposit by the conventional means of sinking a shaft to the deposit. Sometime during this period, Herman Frasch, the first research director of the Standard Oil Company, learned of the Louisiana deposit and devised an ingenious system for mining the sulphur. In 1890, Frasch applied for three patents on his now famous hot-water process for mining sulphur, and in the same year he entered into an agreement to develop this process with Frank Rockefeller, brother of John D. Rockefeller, and F. B. Squires, Secretary of the Standard Oil Company. Arrangements were made to obtain the mining rights to the Calcasieu deposit, and late in 1894 the first Frasch sulphur mine was begun. By 1896, the venture showed sufficient promise to cause the parties to incorporate the Union Sulphur Company which acquired title to the land and mineral rights at the sulphur mine and to the Frasch patents. After several delays in bringing the mine into profitable production, output of Frasch sulphur progressed steadily, rising from less than 25,000 tons in 1903 to an average of over 200,000 tons per year between 1905 and 1911.

The expiration of the original Frasch patents in 1908 stimulated competition in the American sulphur market. As early as 1906, the Gulf Development Company had been formed to prospect for sulphur on the Bryan Heights Salt Dome (often called Bryan Mound) in Brazoria County, Texas. In 1912, the Freeport Sulphur Company was incorporated in Texas to mine sulphur from this property using the Frasch process. On 12 November 1912 the first sulphur was mined from Bryan Mound. Efforts by the Union Sulphur Company to contest use of the Frasch process by Freeport on the grounds that such operations infringed on the patent rights of Union were not upheld by the courts.[4]

A second competitor for the Union Sulphur Company developed when Gulf Sulphur Company was organized in 1909 to develop sulphur deposits on the Big Hill Dome, in Matagorda County, Texas. Bernard Baruch, who controlled this firm, delayed development of the deposit because he felt "with both the Union and Freeport sulphur companies in production, there seemed to be no room in the market for a third producer."[5] By 1916, however, the increased demand for sulphur resulting from the war in Europe prompted Baruch to obtain additional capital

4. See *Union v. Freeport*, U.S. District Court, Delaware, In Equity No. 336; and U.S. Circuit Court of Appeals, 3rd Circuit, Nos. 2391 and 2392, October Term, 1918. A summary of the judgment is given in "Frasch Sulphur Mining Process Decision," *The Journal of Industrial and Engineering Chemistry* (April 1919), pp. 374–75.

5. Bernard Baruch, *Baruch, My Own Story* (New York: Henry Holt and Company, 1957), p. 237.

from the Morgan interests, to reorganize the firm as the Texas Gulf Sulphur Company, and to commence development of the property. While the original plan had been to develop the property modestly, the entrance of the United States into the war abruptly changed these plans. The Big Hill mine came on stream early in 1919.

From 1919 to 1924, Texas Gulf and Freeport accounted for between one-half and three-quarters of the total Frasch sulphur output, with Union Sulphur Company providing the remainder. In 1925, Union closed its mine due to depletion, and subsequently—after making several unsuccessful attempts to secure additional reserves and after having sold its aboveground stocks—was forced to leave the industry. Freeport and Texas Gulf were thus the only Frasch producers until the formation of the Duval Texas Sulphur Company in 1928. A fourth company, Jefferson Lake Oil Company (later Jefferson Lake Sulphur Company) entered in 1932. From 1932 until 1966, these four companies controlled virtually the entire output of Frasch sulphur in the United States, with Texas Gulf and Freeport accounting for nearly 90 per cent of total output and Duval and Jefferson Lake sharing the remainder. During this period, there were at least five unsuccessful attempts made to enter the industry.

In response to a growing shortage of sulphur, several firms took steps to enter the Frasch sulphur industry during 1966 and 1967.[6] Union Texas Petroleum Division, a subsidiary of Allied Chemical Corporation, reestablished the original Union mine in Sulphur, Louisiana, beginning production on 18 September 1966. A newly formed company, Phelan Sulphur Company, constructed a $2.5 million plant to produce about 150,000 tons of sulphur per year from the Nash Dome, a mine abandoned in 1956 by Freeport after a little over 150,000 tons had been produced. Hooker Chemical Company built a pilot plant in June 1967 to produce sulphur from Bryan Mound, the initial mine opened by Freeport in 1912 and abandoned in 1935. In 1968, U.S. Oil of Louisiana, Ltd., reopened the Chacahoula Dome. Between 1955 and 1962, Freeport produced about 1.2 million tons from this deposit, which is thought to still hold over 4 million tons.

Table 2 and figure 2 depict the salt domes that have been mined, their location, and the sulphur obtained from each deposit through 1966. It can be seen that in addition to being highly concentrated economically, the Frasch sulphur industry is also very concentrated geographically. Only 27 of the 200-odd salt domes discovered on the Texas and Louisiana Gulf Coasts have been mined, and of these only 10 have produced as much as 5 million tons of sulphur. In 1968, of the 18 mines producing Frasch sulphur in the United States, only three—Grand Isle,

6. See "New Sources of Sulphur Emerge," *Industrial Minerals* (December 1967), pp. 23, 24, 29.

Table 2.   Production of Frasch Sulphur in the United States by Mines and Companies, in Chronological Order, 1895 to 1966

Dome	Company	Opened	Closed	Total output
				long tons
Sulphur Mine, La.	Union	12-27-94	12-23-24	9,412,165
Bryan Mound	Freeport	11-12-12	9-30-35	5,001,068
Gulf (Big Hill)	Texas Gulf	3-19-19	8-10-36	12,349,597
Hoskins Mound	Freeport	3-31-23	5-26-55	10,895,090
Big Creek	Union	3- 6-25	2-24-26	1,450
Palangana	Duval	10-27-28	3-10-35	236,662
Boling Dome	Union	11-14-28	8-30-29	9,164
Boling Dome	Texas Gulf	3-19-29	OPEN	61,118,065
Long Point	Texas Gulf	3-19-30	10-19-38	402,105
Lake Peigneur	Jefferson Lake	10-20-32	6- 7-36	430,811
Grande Ecaille	Freeport	12- 8-33	OPEN	30,885,243
Boling Dome	Duval	3-23-35	4-25-40	571,123
Boling Dome	Baker-Williams	6- 2-35	12-18-35	1,435
Clemens Dome	Jefferson Lake	5- 3-37	12-14-60	2,975,828
Orchard Dome	Duval	1-29-38	OPEN	5,149,215
Long Point	Jefferson Lake	6- 7-46	OPEN	4,551,472
Moss Bluff	Texas Gulf	6-24-48	OPEN	5,081,343
Starks Dome	Jefferson Lake	6-15-51	12-13-60	840,249
Spindletop Mine	Texas Lake	5-12-52	OPEN	6,310,721
Bay Ste. Elaine	Freeport	11-19-52	12-29-59	1,131,204
Damon	Standard Sulphur	11-11-53	4-20-57	139,618
Garden Island Bay	Freeport	11-19-53	OPEN	7,006,991
Nash	Freeport	2- 3-54	11-23-56	153,115
Chacahoula	Freeport	2-25-55	9-28-62	1,199,015
Fannett	Texas Gulf	5- 6-58	OPEN	1,773,737
High Island	United States	3-25-60	2- 8-62	36,708
Grand Isle	Freeport	4-17-60	OPEN	4,466,021
Lake Pelto	Freeport	11-26-60	OPEN	2,474,693
Gulf (Big Hill)	Texas Gulf	10- 8-65	OPEN	107,830
Sulphur Mine, La.	Allied Chemical	9-18-66	OPEN	1,447
Nash	Phelan Sulphur Co.	11- 7-66	OPEN	622

SOURCE: Industry data for 1895-1966 compiled by Freeport Sulphur Company.

NOTE: Several mines were opened after 1966. Hooker Chemical operated Bryan Mound from 1967 to 1968, and U.S. Oil of Louisiana, Ltd., began operations at Chacahoula in 1967. In 1968, Freeport began operations at Caminada, Jefferson Lake at Lake Hermitage, and Texas Gulf at Lake Bully Camp.

Grande Ecaille, and Boling—were capable of an output of one million tons per year.

Development of offshore Frasch mining

One of the most interesting developments in the Frasch sulphur industry was the introduction of offshore Frasch operations in the Gulf of Mexico. This development was directly related to the extension of petroleum exploration and production into the coastal waters off Louisiana

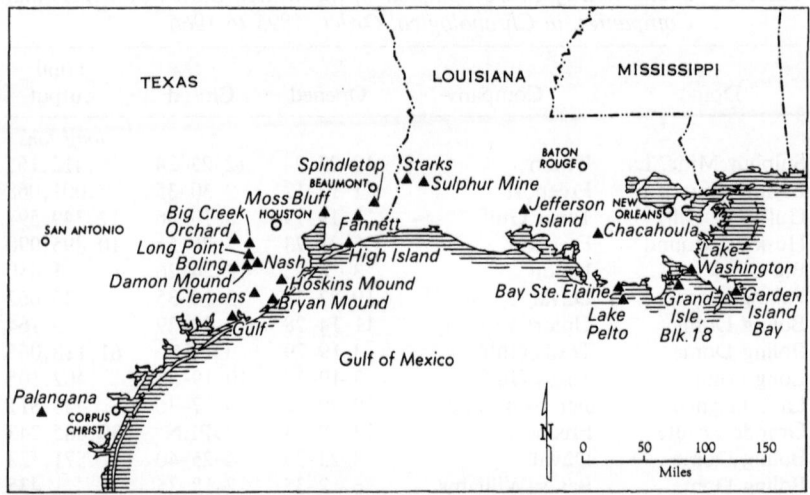

Figure 2.   Frasch sulphur mines on Louisiana and Texas salt domes.

and Texas immediately after World War II. The first offshore deposit of sulphur in the Gulf of Mexico was discovered in 1949 by Humble Oil and Refining Company, while drilling for oil on its Grand Isle, Block 16 lease, located in 50 feet of water, 7½ miles off the coast of Jefferson Parish, Louisiana. In 1954, ten sulphur prospect holes were drilled, and sulphur ore was encountered in eight of them. Humble thus established the existence of a major sulphur deposit with estimated reserves of 30–40 million tons.[7] The sulphur rights at Grand Isle were held under leases originally executed by the State of Louisiana and later confirmed by the federal government pursuant to the provisions of the Outer Continental Shelf Lands Act.

Despite the fact that Humble had obtained authorization from its stockholders in October 1954 to expand its activities beyond oil and gas specifically to develop sulphur, two years later it entered into an agreement with Freeport Sulphur Company which gave Freeport the rights to the Grand Isle sulphur deposit and also the sulphur rights on two nearby domes: Grand Isle, Block 16, where a commercially exploitable deposit had been proved; and West Delta, Block 30, where Freeport was to continue prospecting work begun by Humble.

Construction of the Grand Isle plant began in June 1959 and was completed in 1962. The estimated total cost for this first offshore Frasch process plant was $30 million, of which $8 million was reported to rep-

7. C. O. Lee, Z. W. Bartlett, and R. H. Feierabend, "The Grand Isle Mine," *Mining Engineering* (June 1960), p. 578.

resent the additional costs of offshore operations.[8] In 1968, Freeport placed in operation a second offshore Frasch mine on Grand Isle, Block 16, known as Caminada. The $25 million plant is located in 50 feet of water, 6 miles offshore. It mines a sulphur deposit that lies 1,750 feet under sea level.

Several unsuccessful exploration efforts were undertaken on offshore domes by Texas Gulf and Freeport during the mid-fifties. During the decade following there were no additional attempts to seek new sulphur deposits in the offshore waters. For most of this period, sulphur was in surplus and there was little incentive to undergo the costs of offshore exploration. However, in December 1965, following a turn toward tightness in the sulphur market, the Bureau of Land Management of the U.S. Department of the Interior awarded leases to prospect for sulphur on 72,000 acres of the outer continental shelf off the Texas Coast. The tracts leased are located 40–80 miles offshore from the Galveston-Freeport area and lie under 110 to 180 feet of water. Fifty tracts of 1,440 acres each were leased to seven companies or combines for $33.7 million, an average of $468 per acre. This sum represents the highest price ever paid for mineral lease rights in the outer continental shelf area except for certain oil and gas drainage rights. The leases run for ten years with a royalty payment to the federal government of 10 per cent of the gross product or value of sulphur at the wellhead but not less than $2 per long ton.[9]

High costs are a major obstacle to extensive mining of sulphur offshore. The necessity of securing leases from the federal government by competitive bidding greatly increases the initial costs of offshore as compared with onshore exploration.[10] In addition, the costs of exploratory drilling are much higher offshore, as are the costs of building and operating a Frasch plant. The Grand Isle and Caminada mines are located only 6–8 miles offshore. At both mines, sulphur is shipped via a heated underwater pipeline to shore and then transferred by barge to a storage terminal at Port Sulphur. The tracts leased in 1965, however, are located 40–80 miles offshore. Thus, the problems and costs of shipping

8. *Ibid.* Also, see *The Engineering and Mining Journal* (December 1961), p. 137, for details regarding the expansion authorized in 1961 and completed in 1962.

9. U.S. Department of the Interior, Bureau of Land Management, "Interior Leases Sulphur Rights on the Continental Shelf," *News Release*, 21 December 1965.

10. For example, in 1965, in competitive bidding for offshore leases at San Luis Pass, Freeport paid $1.1 million more than the combined bids entered by the other bidders. Similarly, Texas Gulf Sulphur paid $7.1 million for its leases off Galveston Island, while the combined total of all other bids on these leases amounted to only $385,720! Texas Gulf paid $750,000 each for two leases on which it was the only bidder.

sulphur from mines at such locations to an onshore distribution terminal will be much greater than have been faced to date.

Development of the Mexican Frasch sulphur industry[11]

Petroleum exploration disclosed the existence of sulphur in salt dome formations on the Isthmus of Tehuantepec as early as 1902, but for forty years sulphur was of interest only because it indicated the presence of oil. It remained for General Alfredo Breceda, a hero of the Mexican revolution, to capitalize on the results of the early explorations. In 1942, Breceda and his partner, Manuel Urquidi, succeeded in obtaining the first sulphur concession on the Isthmus from the Mexican Ministry of National Economy.

Under Article 27 of the 1917 Constitution, all minerals in Mexico belong to the people of Mexico, regardless of the ownership of the surface land. Legislation passed in 1930 provided for the establishment of national mining reserves and set up the *Comisión de Fomento Minero* (Mining Development Commission), which was empowered to issue concessions within the reserves for the exploration and development of mineral deposits. Sulphur was not included in the national mineral reserves, and the concessions granted to Breceda and Urquidi were therefore based on a regulation issued in 1941 by the Minister of National Economy. However, this regulation established conditions for sulphur concessions which were similar to those for the concessions granted by the *Comisión de Fomento Minero* (the *Fomento*). Breceda and Urquidi received three concessions of 100 hectares each, the maximum size allowable under the mining law.

In 1943, an *acuerdo*, or decree, of President Manuel Avila Camacho placed all unassigned portions of the Tehuantepec area in a *Zona Azufrosa*, a Sulphur Reserve, under the administration of the *Fomento*. The procedures for obtaining rights of exploration and exploitation for sulphur were thereafter defined by the *Fomento*, rather than by the mining law. However, the Breceda and Urquidi concessions remained outside the control of the *Fomento*.[12]

11. This section is based primarily upon Haynes, *Brimstone: The Stone That Burns*, chaps. 1, 14, and 15. Also, see Robert Sheehan, "The 'Little Mothers' and Pan American Sulphur," *Fortune* (July 1960), pp. 96ff.; John H. Kearney, "A New Empire of Frasch Process Sulphur Is Rising from the Jungles of Mexico," *The Engineering and Mining Journal* (January 1955), pp. 72–77; Dale B. Truett, "Sulphur and the Development of a Chemical Fertilizer Industry in Mexico" (Ph.D. dissertation, The University of Texas, 1967); and Miguel S. Wionczek, "Foreign-Owned Export-Oriented Enclave in a Rapidly Industrializing Economy: Sulphur Mining in Mexico," in *El Nacionalismo mexicano y la inversión extranjera* (Siglo XXI Editores, Mexico, 1967).

12. Wionczek, *op. cit.*, states that Breceda suggested this action, the purpose of which was to complicate future possible direct deals between the authorities and foreign companies, namely the two major U.S. Frasch producers, Texas Gulf and Freeport.

Breceda and Urquidi attempted to secure domestic financial backing during the mid-forties. In particular, they sought to join forces with the *Nacional Financiera* (NAFIN), the national industrial development bank. Surface examinations of areas adjacent to their concessions prompted Breceda and Urquidi to apply for additional concessions; however, they withdrew their request when the government indicated its wish to cede these areas to the *Fomento* for exploitation directed by NAFIN and contracted to the Freeport Sulphur Company. The project reportedly fell through when Freeport demanded a contract in which taxes would be frozen at a low level.[13] Subsequently, the President issued a decree (28 April 1944) which directed NAFIN to organize a corporation to develop the Isthmus sulphur deposits. Breceda and Urquidi were furnished with an office by NAFIN from which they organized exploration work under the name Azufres de Tehuantepec. They entered into an agreement with a trio of U.S. wildcat drillers, the Brady brothers, to conduct an exploration program on their initial concessions. When sulphur in commercially minable quantities was proven to exist, Breceda and Urquidi reaffirmed their intention of joining with NAFIN, on NAFIN's terms, to form a national sulphur industry. However, at this point, NAFIN abruptly decided that it was no longer interested in organizing a sulphur operation on the Isthmus.[14] After rejection by NAFIN and several unsuccessful attempts to obtain domestic financing, Breceda and Urquidi assigned their concessions to the Bradys in exchange for a royalty agreement. The Bradys continued exploration on the Isthmus, and their discoveries subsequently led to the formation of three U.S. corporations, each with a subsidiary incorporated in Mexico, to mine sulphur by the Frasch process.

In 1946, the Mexican Gulf Sulphur Company was formed to mine the San Cristobal concession. The plant, financed in part by a loan from the Export-Import Bank, came on stream in 1954, but returns from the wells were irregular, and the output was about half the expected volume. After efforts to expand output and to discover additional reserves proved unsuccessful, Mexican Gulf's plant and equipment were purchased by the *Nacional Financiera* in 1958.

The Mexican Gulf experience points up the uncertain nature of Mexican Frasch operations as compared with the U.S. Gulf Coast. The Texas-Louisiana mines are true salt dome formations, whereas the Isthmus sulphur formations are a combination of domes and anticlines with a salt core apex. The anticlines, folds or ridges of stratified rock, are a less regular formation than a salt plug, and can present very

---

13. Antonio Acevedo Escobedo, *El azufre en Mexico* (Mexico, D.F.: Editorial Cultura, 1956), p. 185. Also see Wionczek, *op. cit.*
14. It was later established that lack of funds and opposition from U.S. sulphur interests were responsible for NAFIN's withdrawal. Acevedo, *op. cit.*, pp. 188–89.

favorable or almost impossible structures for the operation of the Frasch process. The San Cristobal deposit appeared to be particularly spotty with some very favorable showing mixed with barren holes. Later, other sulphur operations in Mexico were to experience similar difficulties.

The second company established to mine Frasch sulphur on one of the Isthmus concessions, Pan American Sulphur Company (PASCO), fared much better than Mexican Gulf. PASCO was formed in 1947 by a group of Texas oilmen to mine the Jaltipan, Portrerillos, and Teterete domes. PASCO's owners were not initially interested in becoming sulphur producers. Having proved the existence of commercial deposits, they sought to sell their concessions.[15] Only after their attempts to interest Freeport Sulphur Company and Texas Gulf Sulphur Company in purchasing the property were unsuccessful, did PASCO's owners decide to build a plant themselves.[16]

After receiving over $4 million in construction loans from the Export-Import Bank, PASCO brought its plant on stream on 24 September 1954. From the beginning, production proceeded without difficulty. Within eighteen months, the capacity of the plant was expanded by 50 per cent. By 10 December 1956, the company had produced one million tons of sulphur. In the spring of 1955, the Bradys disposed of their stock in PASCO for an estimated $7 million. They retained their royalty interest of $1.00 per ton on sulphur produced from the initial concession and 50 cents per ton on sulphur produced from the concessions acquired in 1950.

In 1951, the Bradys formed a third company, Gulf Sulphur Company, to mine the remaining Breceda-Urquidi concessions on the Mezquital, Soledad, and Salinas domes and part of the Vista Hermosa dome. The company's plant and sulphur mining facilities came on stream on 3 May 1956. Just prior to the commencement of sulphur production from the Salinas dome, Hudson Engineering Company and Baer, Sterns, and Company purchased all the Bradys' Gulf Sulphur stock. Production from the Salinas mine rose slowly from 110,841 tons in 1956 to 373,000 tons in 1962.

Texas Gulf Sulphur Company (TGS) had indicated interest in the sulphur deposits of the Isthmus as early as 1941, and in the fall of 1948 had sent a team of geologists to the Isthmus to survey the deposits there.

---

15. Drilling undertaken in 1944 and 1945 by the Brady brothers proved that the original concessions obtained by Breceda and Urquidi covered only the fringe of the key salt domes. Subsequently, Breceda and Urquidi, acting on behalf of PASCO, were able to obtain by presidential decree additional concessions of 12,217 hectares in adjacent areas. The areas in question were taken out of the national sulphur reserve. The political factors involved in acquiring these additions to the initial concessions are considered in detail by Wionczek, *op. cit.*

16. Sheehan, *op. cit.*, p. 199.

In 1949, the *Fomento* signed the first direct exploitation contract for sulphur with Compañía Exploradora del Istmo, S.A., the Mexican subsidiary of Texas Gulf.[17] The concessions granted Texas Gulf included the Nopalapa dome and portions of the Texistepec dome. The provisions of the contract were not at all in accord with the regulations applicable to the sulphur concessions outside the *Fomento* area. In 1949, both PASCO and Gulf Sulphur were subject to a 6 per cent royalty, a 15 per cent export tax, and a one per cent production tax; nothing was stipulated regarding a tax or royalty ceiling. The TGS contract, on the other hand, provided for a total tax and royalty load of 16 per cent during the first five years of operation, and a maximum of 21 per cent thereafter for the life of the contract (up to forty years). Faced with the prospect of such an inequitable tax, PASCO again drafted the services of Breceda and Urquidi. These two gentlemen succeeded in obtaining equal tax and royalty status for all sulphur producers by virtue of a Presidential Decree of 20 June 1950.[18]

Texas Gulf conducted a thorough exploration of their concession before deciding to construct a plant on the Nopalapa dome. The plant was not brought on stream until 8 February 1957. Production totaled 115,000 tons in 1957 and 1958 and 120,000 tons in 1959, all of which was stockpiled. In February 1960, production was suspended at the Nopalapa mine. Suspension of operations was attributed to the high costs of the operation, as compared with Texas Gulf's domestic mines, and the sharp reduction in sulphur prices throughout the world in the period from 1958 through 1962. During 1967, with sulphur in tight supply and prices rising, Texas Gulf announced plans to place the Nopalapa mine back into production.

In 1953, the Texas International Sulphur Company was formed to mine a surface deposit of sulphur in the State of Baja California. In 1956, the company purchased Central Minera, a Mexican firm which

---

17. Wionczek, *op. cit.*, describes the controversy that raged in Mexico over the opening of the Mexican Frasch sulphur industry to the "international sulphur trusts." Much of this opposition was led by the Breceda-Urquidi interests. Wionczek concludes that this resistance backfired. Not only did it not prevent the granting of the concession to Texas Gulf, but more importantly, it called attention to the presence of a foreign sulphur company in the country and strengthened the forces within the country that were pressing for the closer control of the mining sector.

18. Truett, *op. cit.*, pp. 87–90. In addition to the discrepancy in royalty tax arrangements between the TGS contract and the regulations applying to other concessionaires, other discrepancies existed between the two documents, particularly in terms of the investment required to maintain the concessions (or contract) and the required exploration program. Truett, *op. cit.*, pp. 92–94, and Acevedo, *op. cit.*, p. 118, contend that TGS appeared to have negotiated more with the view of blocking the sulphur mining efforts of potential competitors than of developing its own mining facilities.

had obtained a *Fomento* sulphur concession. Texas International installed a Frasch plant in the Texistepec concession in late 1959. No production was recorded for 1960, and production totaled only 50,000 tons in 1961. In 1962, shortage of capital forced the company to suspend operations, and no production has been recorded since then.[19]

Thus, of the five attempts to mine sulphur in Mexico, two were failures, and a third was discontinued. Of the two that remained, Pan American Sulphur Company was by far the more successful. Table 3 shows the rapid growth of Frasch sulphur output in Mexico, and PASCO's dominant position in the industry.

Table 3.   Output of Frasch Sulphur in Mexico, 1954–65

Year	Total output	PASCO output	Other output	PASCO share of total output
	(............ *thousand long tons* ....)			*per cent*
1954	87.2	31.6	55.6	36.2
1955	475.5	391.8	83.7	82.4
1956	754.1	623.8	130.3	82.7
1957	993.4	723.1	270.3	72.8
1958	1,201.4	822.0	379.4	68.4
1959	1,266.8	886.9	379.9	70.0
1960	1,252.3	1,027.6	224.7	82.0
1961	1,171.8	882.7	289.1	75.3
1962	1,335.7	982.9	352.8*	73.6
1963	1,448.0	1,121.2	326.8*	77.4
1964	1,625.5	1,275.8	349.7*	78.5
1965	1,519.6	1,142.6	377.0*	75.2

SOURCE: Secretaría de Industria y Comercio, Dirección General de Estadística.
NOTE: All figures converted from metric tons to long tons. Conversion factor used = 0.98419.
* After 1962, Gulf Sulphur Company.

Relations between the Frasch producers and the Mexican government became increasingly strained after the passage of a new mining law in 1961 which required firms conducting mining operations in Mexico to be owned at least two-thirds by Mexican nationals. Pressures were exerted on both Pan American and Gulf Sulphur to mexicanize their operations. These pressures culminated in 1965, when the Mexican government abruptly imposed export controls on the Frasch sulphur industry. The stated objective of this action was to bring increases in exports into line with expansions in reserves. However, two equally important objectives which received little public attention were to encourage the development of a domestic chemical fertilizer complex,

19. *Sulphur* (June 1963), p. 17.

and to bring Frasch sulphur operations on the Isthmus into agreement with the provisions of the 1961 mining law.

As might have been expected, the initial reaction of the American financial press to the action taken by the Mexican government was quite hostile. One writer stated:

> For a quarter-century Mexico has sought, with considerable success of late, to live down its ill-advised expropriations of U.S. and British oil interests. Last month the government suddenly acted like a reformed bandit who can not give up his old tricks. Robbing the gringo, as everyone should have learned by now, is no way to get rich.[20]

The controls on sulphur exports proved much less restrictive than had been anticipated. Not only did the export authorizations in 1965 and 1966 not fall below previous levels, but they provided for a steady expansion of exports beginning in 1967 following the establishment of sizable new reserves. Without a doubt, the Mexican policy was highly successful in bringing about both an expansion in reserves and a program for the development of a chemical fertilizer complex in Mexico using Mexican sulphur.[21] At the same time, by demonstrating its power and will to influence the operations of the major Frasch producer, the Mexican government succeeded in bringing closer the mexicanization of the Isthmus sulphur industry. Late in 1966, negotiations proceeded to permit purchase by Mexican financial interests of a two-thirds ownership in Pan American Sulphur Company. Subsequently, PASCO was brought within the framework of the 1961 mining law.

The attitude of the Mexican government toward exploration and mining of sulphur by foreign firms was aptly summarized in April 1965 in a statement by the then-Secretary of National Properties, Alfonso Corona del Rosal:

> Concessions for sulphur exploration will be given to some companies that have sought them, within the framework of the present Mining Law; that is, to companies with at least 66 per cent Mexican capital, and preferably to those who promise to convert the sulphur into fertilizers for the demands of our agriculture and for export of the excess. Those companies will have the right, as an incentive to their explorations, to export 10 per cent of the sulphur reserves they discover.[22]

20. "Robbing the Gringo," *Barron's* (3 May 1965), p. 1.

21. Pan American, for example, spent $1 million in 1965 and 1966 in drilling 100 exploratory wells. They also entered into an agreement with two U.S. investment houses to establish a firm, 49 per cent owned by U.S. interests and 51 per cent owned by Mexican nationals, to build a $44 million fertilizer complex scheduled to commence production in late 1968. See "Pan American Sulphur Says '65 Profit, Sales Will Trail Last Year," *Wall Street Journal* (28 December 1965), p. 14.

22. Cited in "Scramble on for Mexican Sulphur," *The Engineering and Mining Journal* (May 1967), p. 105.

This policy, which appears to be still in effect, is a compromise that gives each of the interested parties significant advantages at minimum cost. The Mexican government gains control over the sulphur industry and the accompanying political advantages. Foreign-owned producers gain a fair return on their investment and the advantage of lower taxes on their operation. Finally, Mexican investors are given an opportunity to invest in the domestic sulphur industry.

Recovered Sulphur

When the practice of removing hydrogen sulphide from natural and manufactured gases was begun late in the nineteenth century, the hydrogen sulphide was treated as a waste and either vented to the atmosphere or burnt in a flare or under a boiler. Later, with the appearance of anti-pollution and conservation laws and the increasing use of gases with high concentrations of hydrogen sulphide, it became necessary to convert the hydrogen sulphide into either elemental sulphur or sulphuric acid. The shortage of sulphur in the early 1950s provided an economic incentive to accelerated recovery of sulphur from acid gas streams, and since 1954 sour natural and refinery gases have become a major source of sulphur.

Hydrogen sulphide can be separated from sour natural or refinery gases in several ways, but the processes most widely used today are ones such as the Girbotol process that recover hydrogen sulphide as a gas. However, as there is only a limited industrial demand for hydrogen sulphide, the gas is usually converted into a more marketable form of sulphur. Conversion into elemental sulphur is preferred to conversion into sulphuric acid because the process is simpler and the product is easier to store, handle, and transport. Today, most conversion plants use a method based on the Claus-Chance process to convert a concentrated gas stream containing hydrogen sulphide into sulphur dioxide and then into elemental sulphur.[23] The basic flow scheme of a Claus-Chance plant is shown in figure 3.

First, hydrogen sulphide-bearing gas is burned in a specially designed reaction furnace with a controlled amount of air, forming hydrogen sulphide. In some plants, as much as two-thirds of the hydrogen sulphide feed is bypassed around the furnace, in which case the stream passing through the furnace is burned all the way to sulphur dioxide. The gas passes from the reaction furnace, through a waste-heat boiler, and then

---

23. See James W. Estep, Guy T. McBride, Jr., and James R. West, "The Recovery of Sulphur from Sour Natural and Refinery Gases," in *Advances in Petroleum Chemistry and Refining*, vol. 6, ed. John J. McKetta, Jr. (New York: Interscience Publishers, 1962), p. 329.

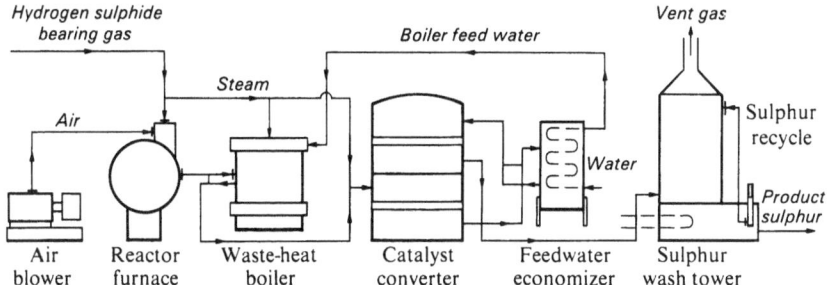

Figure 3. Flow scheme of sulphur recovery plant.

enters the catalyst converter. The hot gas from the first catalyst bed is cooled in a boiler feedwater economizer and then passes through a second catalyst bed which is operated at a lower temperature to give maximum conversion. Gas from the second catalyst chamber passes through a sulphur wash tower where condensation and removal of elemental sulphur occurs. The residual gas is vented to the atmosphere, while product sulphur is taken off to storage. With this process, up to 90 per cent of the sulphur may be recovered.

Domestic production of recovered sulphur

The discovery of several sour natural gas fields in southwest Arkansas in the early 1940s brought into existence the first commercial operation in the United States to recover elemental sulphur from sour gas streams. The Arkansas gas contained about 8 per cent hydrogen sulphide and an equal concentration of carbon dioxide. In 1941, the Southern Acid and Sulphur Company, now a part of Olin Mathieson Chemical Corporation, engaged the Ohio State University Research Foundation to develop in the laboratory a process for the recovery of sulphur from the acid gas separated from Arkansas sour natural gas, and in 1942 the company built a pilot sulphur recovery plant at Magnolia, Arkansas. About the same time, Susearch, an affiliate of the Texas Gulf Sulphur Company, built a pilot plant at McKamie, Arkansas, to conduct an independent study of the utilization of hydrogen sulphide derived from sour natural gas for the production of elemental sulphur. These pioneer efforts were instrumental in the commercial development of sulphur recovery from sour gas streams.

In 1944, Southern Acid constructed the first commercial sulphur recovery plant at McKamie, Arkansas. In 1949, sulphur recovery was begun from sour natural gas produced in the Worland-Manderson-Powell area of north-central Wyoming. The first company to recover sulphur from the natural gas of this region, which contained up to 30 per cent hydrogen sulphide, was the Stanolind Oil Company (later the Pan American Petroleum Corporation), which built a plant with a capacity of 75 tons per day at Elk Basin, Wyoming, in 1949.[24] In the 1950s, sulphur recovery operations were begun in the Permian Basin area of West Texas and southwestern New Mexico. The first sulphur recovery plant in this field was the 20-ton-per-day plant built by the Sid Richardson and Odessa Natural Gas companies at Odessa, Texas, in 1952.[25] The most recent discoveries of sour natural gas have been made in central Texas in the Smackover Trend, a geological formation extending from Mexico through central Texas to as far east as Alabama. Tide Water Oil Company built the first plant to recover sulphur from gas produced in this field in 1960 at Mt. Vernon, Texas.[26]

In each of the major sour natural gas fields, the first recovery plants were soon followed by additional plants having recovery capacities ranging from 10 tons to 300 tons per day. The only sulphur recovery plant in the United States having a capacity in excess of 300 tons per day is the Northeast Edgewood Field plant of Pan American Petroleum Corporation in Van Zandt County, Texas. The plant, built to process sour gas from the Smackover Trend, has a capacity of 900 tons of sulphur per day and was completed in 1964.[27]

Corresponding to the development of sulphur recovery from sour natural gas has been the recovery of sulphur at oil refineries processing sour crudes. The first large-scale sulphur recovery unit at an oil refinery in the United States was installed at Watson, California, by Hancock Chemical Company in 1949.[28] Since that time, additional recovery plants have come on stream at refineries in California and Texas and along the Eastern Seaboard, and sulphur is now being recovered at forty-nine refineries in the United States.

Factors that encourage sulphur recovery at oil refineries include: (1) laws controlling air pollution; (2) use of crudes with a high sulphur content; (3) demand for desulphurized products; (4) value of sulphur produced; and (5) use of sulphuric acid for oil refining purposes. Factors (1) and (3) have been particularly important in densely populated and highly developed areas. Thus, California, where stringent regula-

24. F. J. Kelly, "Sulphur Production and Consumption in Eight Western States," U.S. Department of the Interior, Bureau of Mines, *Information Circular 8094*, 1962, p. 72.

25. H. H. Jones and R. A. Graff, "West Texas' First Sulphur Recovery Unit," *The Oil and Gas Journal* (21 April 1952), pp. 122–23 and 153–54.

26. Leonard P. Larson and Victoria M. Roman, "Sulphur and Pyrites," U.S. Department of the Interior, Bureau of Mines, *Minerals Yearbook*, 1960, vol. 1.

27. See "New Sour Gas Plant Planned in Texas by Pan American," *The Oil Daily* (25 June 1963), and Robert B. Bizal, "Industry Adding 48 New Plants in '64," *The Oil and Gas Journal* (16 March 1964), p. 142.

28. See "U.S. Recovered Sulphur Capacity Shows New Growth Rate Peak in 1965," *Sulphur* (April/May 1966), pp. 13–16.

tions controlling air pollution exist and practically all hydrogen sulphide available from oil refineries is converted into either sulphur or sulphuric acid, accounts for over 24 per cent of total capacity for sulphur recovery from oil refineries in the United States.[29] Factor (2) is becoming increasingly important as larger amounts of sour crude are being refined. The importance of factor (4) will depend upon the distance of the oil refinery from sulphur markets and the state of the sulphur market. Undoubtedly, the shortage of sulphur in this country in the early fifties led to increased recovery of sulphur in the latter half of the decade. While factor (5) is not of major significance in the United States, it is important where refineries are located near the centers of oil production, such as Abadan or in the Caribbean.

At present, the forty-eight plants producing natural gas account for about 57 per cent of domestic recovered sulphur production. With the possible exception of the recently constructed plant in Van Zandt County, Texas, sulphur recovery facilities in the United States have been too small and too far removed from the major markets to have an appreciable impact on the total sulphur market. Only a few plants produce more than 50,000 tons per year, and only two firms, Pan American Petroleum Corporation and Getty Oil Company, are capable of producing as much as 100,000 tons of recovered sulphur per year.

Foreign production of recovered sulphur

While domestic production of recovered sulphur has had little impact on domestic Frasch sulphur markets, foreign production of recovered sulphur has made a major impact on world sulphur markets and, in particular, on U.S. exports of sulphur. Recovered sulphur is of major importance in Canada and France, and it is becoming increasingly important in the Middle East.

Canada. The existence of natural gas in Western Canada had been known since before the turn of the century; however, it was not until the late 1940s that oil discoveries in Alberta prompted extensive exploration for oil in the western provinces which brought to light extensive gas deposits.[30] The commercial development of natural gas in Western Canada was delayed until two conditions could be met: first, ample reserves to satisfy both domestic and export demands over a long period had to be proved; and second, the approval of both export agencies in Canada and import agencies in the United States had to be obtained and satisfactory long-term contracts negotiated. These conditions were not satisfied until the late fifties. Thus, the major growth in gas processing

29. See "Sulphur Recovery at Oil Refineries," *Sulphur* (February 1963), pp. 27–28.
30. See Eric J. Hansen, *Dynamic Decade* (Toronto: McClelland and Stewart, Ltd., 1958), pp. 223–25.

in Western Canada occurred after two major pipelines were brought on stream in August 1957—the Westcoast Transmission Company pipeline from the Peace River area of northern Alberta to Vancouver and the Pacific Northwest and the Trans-Canada Pipeline Company pipeline from central Alberta to eastern Canada and northeastern United States. In 1961, lines were extended to Montana and the midwestern section of the United States, and in 1962, a line built by Alberta Gas Trunk Line brought Canadian gas into California. The growth in demand for Canadian natural gas which attended the expansion of pipeline facilities is shown in table 4.

Table 4.   Demand for Canadian Natural Gas, by Markets, 1960–63

million cubic feet per day

Markets	1960	1961	1962	1963
Domestic sales	886	1,000	1,100	1,300
United States sales:				
Pacific Northwest				
Via Westcoast Line	0	233	270	275
Via Alberta Natural Line	0	0	120	140
California	0	0	320	370
Montana	0	47	75	75
Midwest	0	166	179	170
Total U.S. sales	0	446	964	1,030
Total sales	886	1,446	2,064	2,330

SOURCE: *Oilweek* (27 January 1964), p. 23.

By the time pipelines were completed, demands for Canadian natural gas had risen to such an extent that it was necessary to tap sour gases to fulfill the contracts. The production of gas containing sizable amounts of hydrogen sulphide marked the birth of the Canadian recovered sulphur industry. The hydrogen sulphide content of natural gas being produced in the prairie provinces of Canada ranges from 1–2 per cent to as high as 30–40 per cent. If the hydrogen sulphide content of the raw gas stream is under 10 per cent, sulphur is generally looked upon as a by-product of natural gas processing. At hydrogen sulphide concentrations ranging from 10 to 20 per cent, sulphur and natural gas are considered coproducts. Where the hydrogen sulphide content of the crude gas stream is in excess of 20 per cent, the natural gas operation becomes a by-product of sulphur recovery operations. These classifications are arbitrary, and the divisions between by-product, coproduct, and main product status are affected by the relative prices of the two commodities.

A plant can vary the hydrogen sulphide content of its gas intake by drawing crude gas from more than one pool, or from several zones within a pool. By drawing on the sourer pools or zones, the plant operator can increase the output of sulphur without affecting the total gas throughput of the plant. In addition, where the hydrogen sulphide content of the raw gas stream is extremely high, it is possible to process the gas to recover the contained sulphur and recycle the gas back into the formation. At least one operation of this type is in operation in Canada today.

In 1953, only three natural gas processing plants were in operation in Western Canada. By 1956, the number of gas processing plants had grown to nine. The number of plants increased steadily after 1956, as did sulphur recovery from natural gas operations, and by the end of 1968, twenty-four sulphur recovery plants were in operation in Western Canada. The location of plants, the controlling companies, and recovery capacities are given in table 5. It should be noted that while these facilities have the capacity to produce over 3.4 million tons of sulphur per year, actual production of recovered sulphur in 1968 was only 2.6 million tons. Output will rarely if ever approach capacity because the plants are designed primarily to process gas, and gas sales contracts thus control sulphur production. Plants must be capable of meeting peak gas demands but will operate at such rates only briefly. Additional capacity under construction in 1968 and due for operation in 1969 will bring total annual capacity to nearly 5 million tons.[31]

France. Sulphur recovery from sour natural gas in France followed the discovery of a large gas field in the Lacq oilfield in southwestern France by the Société Nationale de Pétroles d'Aquitaine (SNPA) in December 1951. The hydrogen sulphide content of the gas was 15.3 per cent, and for nearly four years the French steel industry wrestled with the problem of engineering a metal alloy that could cope with the extensive corrosion that results from processing natural gas with such a high percentage of hydrogen sulphide. In 1955, the problem was finally solved with the development of a chrome-aluminum-molybdenum alloy.

Recoverable reserves in excess of 7 trillion cubic feet of sour gas having been proved, a Girbotol-type sulphur recovery plant was designed by the Ralph M. Parsons Company, and a multi-stage expansion program was adopted. The first unit, a 200-ton-per-day plant was completed in 1957. By the end of 1960, five units had been completed bringing total plant capacity to 4,100 tons per day, or about 1.5 million tons per year. Production of recovered sulphur at Lacq in 1968 was 1.575 million tons.

In 1965, SNPA discovered a sizable field of sour natural gas at

31. Letter dated 5 May 1969 from C. M. Bartley, Mineral Processing Division, Mines Branch, Department of Energy, Mines and Resources, Ottawa, Canada, to the author.

Table 5.   Sulphur Recovery Plants in Western Canada, 1968

Operating company		Location (Alberta unless specified)	H₂S	Annual plant capacity	1968 production
			per cent	(..... 1,000 long tons .....)	
Amerada Petroleum	1.	Olds	11	63.0	55.5
Athabasca Oil	2.	Mildred Lake	n.a.	105.0	23.5
Canadian Delhi Oil	3.	Minnehik-Buck Lake	n.a.	6.3*	5.1
Canadian Fina Oil	4.	Wildcat Hills	4	50.0	42.5
Canadian Superior	5.	Harmattan-Elkton	53	300.0	240.3
Gulf Canadian	6.	Nevis	3–7	45.0*	42.1
	7.	Pincher Creek	10	114.0	87.0
	8.	Rimbey	1–3	120.0	94.7
Home Oil	9.	Carstairs	1	16.0	11.3
Hudson's Bay Oil and Gas	10.	Edson	3	83.0	55.4
	11.	Lone Pine Creek	8–17	38.3*	23.9
	12.	Caroline	n.a.	6.3	—
	13.	Kabob South	n.a.	365.4	—
	14.	Fox Creek	n.a.	350.0	—
Imperial Oil	15.	Redwater	3	4.5	2.0
Jefferson Lake Petrochemicals	16.	Taylor Flats, B.C.	3	130.0	48.8
	17.	Savannah Creek	13	140.0	26.6
Pan American Petroleum	18.	E. Crossfield	38	518.0	—
	19.	Bigstone	n.a.	110.0	—
Petrogas Processing	20.	E. Calgary	31	700.0*	600.4
Royalite Oil Company	21.	Turner Valley	4	10.5	5.7
Shell Canada	22.	Jumping Pound	3–5	87.5*	55.2
	23.	Innisfail	14	40.0	20.6
	24.	Waterton	18–25	576.5	504.7
Standard Oil of California	25.	Nevis	7	50.0*	44.3
Socony Mobil	26.	Wimbourne	14	85.4	69.0
Steelman Gas	27.	Steelman, Sask.	1	4.0	3.7
Texas Gulf Sulphur	28.	Okotoks	33	150.0	130.4
	29.	Windfall	16	495.0*	367.4
	30.	Red Deer River	n.a.	175.0	—
Banff Oil	31.	Rainbow Lake	n.a.	24.5	—
Total				3,414.0	2,560.1

SOURCE: Letter of 5 May 1969 to the author from C. M. Bartley, Mineral Processing Division, Mines Branch, Department of Energy, Mines and Resources, Ottawa, Canada.
* Indicates expansions planned for 1969. Total expansions estimated to equal 175,400 long tons.
n.a.—Not available.

Meillon, 15 miles southeast of Lacq, with a hydrogen sulphide content of 5–6 per cent. A 100,000–150,000 ton recovery unit was completed and placed in operation in 1967.

Middle East. The Middle East is also becoming an important source for sour gas and petroleum sulphur recovery. Within the past few years, several projects tied to fertilizer and petrochemical complexes have been announced. The largest of these is the $100 million sour gas plant being built in Iran by Allied Chemical Corporation and the National Petroleum Company, a state-owned firm. The complex will have an initial daily capacity of 1,000 tons and an eventual one of 1,500 tons.[32]

32. *Chemical Week* (2 March 1968), p. 46.

Additional large plants being planned for Saudi Arabia, Kuwait, and Iraq show promise of expanding Middle East output of recovered sulphur in the next few decades.[33]

Growth since 1950

The increase in production of elemental sulphur from sour natural and refinery gases in the United States, Canada, and France for the period since 1950 is shown in table 6. While production of recovered sulphur has shown a continuous but moderate growth in the United States, its growth has been quite rapid in Canada and France. Taken together, recovered sulphur sources in the United States, Canada, and France now account for about 23 per cent of total world sulphur production (excluding Communist nations), and other recovered sulphur sources account for an additional 2 per cent. It appears likely that recovered sulphur will supply an increasing share of the world's sulphur requirements in the future. This possibility is discussed in chapter 8.

Table 6.   Production of Recovered Sulphur in the United States, Canada, and France, 1950–67

1,000 long tons

Year	United States	Canada*	France	Total
1950	142	0	0	142
1951	184	0	0	184
1952	252	4	0	256
1953	342	16	0	358
1954	359	19	0	378
1955	399	26	0	425
1956	465	30	0	495
1957	511	96	28	635
1958	640	166	127	933
1959	686	263	419	1,368
1960	767	245	778	1,790
1961	858	352	1,080	2,290
1962	900	621	1,326	2,847
1963	947	1,116	1,386	3,449
1964	1,021	1,597	1,487	4,105
1965	1,215	1,847	1,497	4,559
1966	1,240	1,823	1,516	4,579
1967	1,268	2,073	1,719	5,060

SOURCE: U.S. Department of the Interior, Bureau of Mines, *Minerals Yearbook* for applicable years.

* Sales figures (1952–67).

33. See "New Sources of Sulphur Emerge," *Industrial Minerals* (December 1967), p. 29; and *The Engineering and Mining Journal* (February 1968), p. 97.

Chapter 3

Frasch Sulphur Costs of Production
and Conditions of Entry

The supply function for an industry's product is influenced by the nature of costs and the conditions of entry, as well as by the size and distribution of sellers. This chapter contains an analysis of the costs of producing Frasch sulphur, followed by an examination of the conditions of entry into the industry and the general supply conditions for Frasch sulphur.

Costs of Producing Frasch Sulphur

In the short run, the number of plants and firms that can produce Frasch sulphur is relatively fixed. Costs per unit of output for existing plants and firms are dependent upon the rate of output; the prices of input factors; the production function, i.e., the physical productivity or efficiency of management, labor, and other factors which combine to produce output; and selling expenditures. In the long run, however, both the number of plants and firms in the industry and the scale of each plant and firm can be varied. The long-run supply curve for Frasch sulphur is thus related to the number of firms in the industry, the size or scale of each plant or firm, and the resulting marginal costs for these firms.

Technological considerations

When sulphur is mined by the Frasch process, the basic operations are: (1) drilling production wells into the deposit, and (2) supplying hot water for use in melting the sulphur contained in the deposit. The wells have a life of from a few days to two years, with about one year being average, and about 35,000 long tons can be extracted from an average sulphur well before the cavity becomes too large for effective production. The economics of a given deposit, however, depends primarily on the water ratio, i.e., the number of gallons of hot water required to mine one ton of sulphur, because the water ratio determines the maximum production rate that can be sustained from a given boiler plant.[1]

The water ratio itself is determined by the physical characteristics of the deposit. A typical formation of 25 per cent sulphur, 65 per cent rock, and 10 per cent water theoretically should require about 1,185 gallons of 330° F. water per ton of sulphur extracted. However, if the limestone formation containing the sulphur is too porous, much of the water pumped into the formation will be lost through leakage, and the amount of sulphur produced relative to the water input may be small. On the other hand, if the deposit is too tight, the water will be unable to circulate through the formation and the amount of sulphur obtained from each well will be small. Actual water ratios range from as little as 1,000 gallons to as much as 12,000 gallons per ton of sulphur recovered.[2] In general, it would appear that the water ratio varies inversely with the amount of recoverable sulphur in the deposit. Thus, water ratios are considerably lower for domes capable of producing 10 million tons over their life than for deposits with a yield of only one-half million tons.

In addition to varying between deposits, the water ratio varies over the life of the deposit and over the life of any single well. For any dome, the water ratio is an average of the ratios in each well. When a well is first drilled, the water ratio will be high because of the time required for the deposit to heat up. As production from a well continues, the water ratio usually falls until the minimum level of production is reached. Towards the end of the well's life the water ratio begins to increase as the outer limit of the area through which water from the well can circulate is reached. It is this increase in the water ratio which signals the end of a well's productive life. Over the life of a dome, the water ratio will also vary. As production wells approach the fringes of the deposit, the water ratio usually rises. If the deposit is highly faulted, the water ratio may vary greatly over the various sections of the deposit.

The size of Frasch plants (in terms of millions of gallons of water per day) varies directly with the size of the deposit. If the deposit is small—500,000 tons per year or less—the plant must be operated at full capacity in order to maintain production and prevent the dome from cooling. Since operating costs are fixed by the capacity of the

1. Because the amount of published technical information on sulphur production by the Frasch process is limited, this section is based largely on information obtained in interviews with officials of Frasch sulphur firms. These interviews are listed in the bibliography.

2. Texas Gulf Sulphur Company, *Modern Sulphur Mining* (rev. ed.; New York: Texas Gulf Sulphur Company, Inc., 1961), p. 7.

boiler plant, the tonnage over which these costs can be spread then depends upon the water ratio experienced. For larger deposits, a firm has an option of altering the rate of mining by varying the number of wells drilled. The larger plants can vary production in response to changes in demand, and generally operate within a range of 60–100 per cent of capacity.

Engineering cost estimates

There are no published cost estimates for Frasch sulphur mining that break down costs into variable and fixed components. Furthermore, the scarcity of Frasch sulphur deposits and the small number of plants and firms tend to make each Frasch operation unique, and it is difficult to make any generalizations regarding the size of plant and costs of production. The diversification of Frasch sulphur firms into production of other mineral products, which has occurred with increasing frequency in recent years, also limits the validity of conclusions regarding sulphur production costs based on statistical analysis of accounting records. On the other hand, the basic nature of all Frasch operations is the same, and the components of plant and equipment are similar to those found in other industries. Therefore, it is possible to derive generally valid engineering cost estimates which can be used to indicate both the range of unit costs over plants of differing size and domes of differing quality, and the behavior of unit costs over operations of a given plant at various levels of capacity.

In this section, investment costs for three Frasch plants of different capacities are first presented. Next, unit costs for these plants are computed, assuming operation at full capacity. Finally, unit costs for a given plant are evaluated over a range of capacity utilization. Where possible, estimates for both investment and operating costs have been compared with actual published figures, and appropriate adjustments have been made.

Estimates of investment costs for three Frasch plants within the normal daily water capacity range are as follows:

	Plant 1	Plant 2	Plant 3
Capacity (1,000 gal. water/day)	8,000	4,000	1,000
Estimated investment cost ($1,000)	7,200	4,720	1,500
Investment/1,000 gal. water capacity ($1,000)	900	1,180	1,500

These estimates were made by pricing the major items of equipment used in a Frasch sulphur plant. It was assumed that the largest plant could be scaled down to successively smaller capacities. The costs include only expenditures for physical plant and equipment for an onshore Frasch operation built in 1960 on the Texas-Louisiana Gulf Coast; they exclude all costs of exploration and development. The results are generally in agreement with costs that have been released by the sulphur companies. For example, the one-million-gallon-per-day plant of U.S. Sulphur Company at High Island was estimated to have cost $1.5 million new.[3] The 1.75-million-gallon-per-day Bay Ste. Elaine plant of Freeport Sulphur Company, built in 1952, cost $2 million. This amounts to $1,143 per 1,000 gallons of daily water capacity, which is below the estimated costs given above, but Freeport disclosed that the investment costs for this plant were below normal.[4]

These plant investment estimates indicate that costs for fixed plant and equipment rise sharply as the size of the plant increases, but that larger plants cost much less per unit of water capacity. Investment costs per thousand gallons of daily water capacity range downward from $1,500 for the one-million-gallon-per-day plant to $900 for the eight-million-gallon-per-day plant. Economies in construction costs tend to level off at the upper end of the range, but investment costs per unit of capacity continue to decline. Thus, the most economically sized Frasch plant to construct, in terms of investment cost per unit of capacity, would have a capacity in excess of eight million gallons per day.

The savings in investment costs per unit of capacity also generate savings in financial charges and depreciation expenses. These demonstrate the same trend as investment costs per unit of capacity. At an assumed 6 per cent, financial charges per thousand gallons of daily water capacity decline from $90 for the plant with a capacity of one million gallons per day to $71 for the plant with a capacity of four million gallons per day. Thereafter, the savings are less pronounced, declining $17 per thousand gallons between the four-million-gallon-per-day plant and the eight-million-gallon-per-day plant. The annual savings on financial charges caused by the $19 differential between the two smaller plants amounts to about $77,000 while the corresponding savings resulting from the $17 differential between the two larger plants totals $136,000. The significance of such savings varies directly with the rate level.

The decline in investment costs per unit of capacity realized in the construction of larger-scale plants also yields corresponding savings in depreciation charges. Assuming straight line depreciation on the basis of a fifteen-year life, the depreciation charges per thousand gallons of daily water capacity decrease from $100 for the one-million-gallon-per-

3. X. T. Stoddard, Scout Memo, "High Island Dome Sulphur Mine, Galveston County, Texas," for the Humble Oil and Refining Company, 22 July 1959, p. 14.
4. K. T. Price, "Freeport Mines Sulphur by Boat," *Engineering and Mining Journal* (December 1952), p. 102.

day plant to $79 and $60 respectively for the four-million and eight-million gallon-per-day plants. The unit depreciation charges thus reflect proportionately the reductions in investment costs per unit realized through the construction of larger scale plants.

A major portion of the total unit costs for full utilization of any single plant is accounted for by cost elements which are, for the most part, variable in nature. Production costs based on the average weighted figures for all three plant scales can be broken down approximately as follows:

	Per cent of total
Variable costs:	75.7
Natural gas	19.8
Treated water	1.8
Operating supplies	0.6
Production wells	22.0
Sulphur loading	11.2
Severance tax	20.3
Fixed costs:	24.3
Maintenance	2.7
Operating labor	5.7
Supervision	1.3
Labor and plant burden	7.0
Property taxes and insurance	1.6
Depreciation	6.0

In general, the larger plants have a somewhat higher percentage of fixed costs than shown above, and the smaller plants have a somewhat lower percentage. The breakdown of costs is an arbitrary one, and semivariable costs such as operating labor and supervision, which do not in fact vary over the normal range of operations, are listed as fixed costs. Production wells, however, are classified as variable costs, even though their average life is one year, because the only way a Frasch producer can vary production on a given deposit is to change the rate of drilling wells. The operators of the larger plants do consider wells as variable costs. For smaller plants operating with only a few wells, production wells are considered as fixed costs. The estimated unit costs given above do not include exploration and development costs, which are fixed, nor do they include royalty payments, which are variable in nature.[5]

In evaluating unit costs, consideration must be given not only to the

5. In general, exploration costs are capitalized and amounts equivalent to such exploration costs, together with the estimated related future federal income tax savings, are charged to income. There is no standard royalty agreement in the industry. Many royalty payments are based on a fixed percentage of net profits from the operation after all costs for development (including the cost of plant and equipment) are fully recovered.

size of the plant but also to the quality of the deposit, as reflected by the water ratio. Table 7 illustrates how each of these factors affects unit production costs. Of the variable costs in table 7, the estimates for natural gas and water costs were based largely on operating data published by the sulphur operators for the various domes. It should be noted that although gas consumption is the major heat requirement, it is not directly proportional to the water ratio experienced. Additional quantities of steam are used for driving the pumps, generating power, and heating sulphur transfer lines. Gas requirements decline slightly with increases in plant scale owing to the greater operating efficiency of larger boiler installations. The gas rate used ($0.23 per thousand cubic feet) is believed typical for the Gulf Coast during the period 1961–62. Production wells are short-lived, averaging about one year, and have been treated as an operating expense. Some companies, however, do capitalize them and recover the capital through depreciation. The wells are shallow (1,500 to 1,800 feet, in general), and on the basis of petroleum drilling costs in South Louisiana each well was estimated to cost $35,000, including the casing and pipe. Since pipe is often recovered and reused, actual well costs probably run under $35,000. Sulphur loading costs were computed on the basis of a flat charge of $0.57 per long ton of sulphur produced. The state severance tax on sulphur in both Texas and Louisiana is $1.03 per long ton, and this tax has been included as a variable cost.

Fixed costs were based largely on the amount of investment. Maintenance costs were estimated at 3 per cent of the investment per year, a rate that fits the nature of the plant—primarily a boiler installation. Since the number of employees engaged in production does not appear to vary with output over the range of likely plant operation, all labor requirements were considered as being fixed in nature. Operating labor was computed on the basis of an hourly wage of $3.00 with a shift requirement of fourteen men for plant 1 and ten men each for plant 2 and plant 3. Supervision was based on an estimate of eight shift foremen at $7,500 per year and a plant superintendent at $10,000 per year. Labor and plant burden were taken as being equal to operating labor and supervision costs. Local property and miscellaneous taxes and insurance costs were estimated to be 1.75 per cent of the investment per year, a rate generally applicable under normal circumstances. Depreciation was estimated on the basis of a fifteen-year life, which is common for some sulphur producers. Other producers have used a unit of production as the basis for depreciation, which leads to unit charges comparable to those shown for full capacity operation, assuming the plant's capacity in terms of units of output does not vary over its life.

The engineering cost estimates given in table 7 are in general agree-

Table 7.   Estimated Unit Production Costs for Frasch Sulphur

Item	Case I. Plants of different capacity with the same water ratio			Case II. Plants of the same capacity with different water ratios		
	Plant 1	Plant 2	Plant 3	Plant 1	Plant 2	Plant 3
Plant investment	$7,200,000	$4,720,000	$1,500,000	$4,720,000	$4,720,000	$4,720,000
Capacity:						
Gallons of water per day	8,000,000	4,000,000	1,000,000	4,000,000	4,000,000	4,000,000
Tons of sulphur per day	5,000	2,500	625	2,500	1,000	444
Tons of sulphur per year	1,750,000	875,000	218,750	875,000	350,000	155,400
Water ratio	1,600	1,600	1,600	1,600	4,000	9,000
Investment per 1,000 gal.	$ 900	$ 1,180	$ 1,500	$ 1,180	$ 1,180	$ 1,180
Investment per ton of sulphur	$ 4.11	$ 5.39	$ 6.85	$ 5.39	$ 13.49	$ 30.37
Variable costs:	$/ton	$/ton	$/ton	$/ton	$/ton	$/ton
Natural gas	1.00	1.01	1.02	1.01	2.53	5.69
Treated water	.09	.09	.09	.09	.23	.51
Operating supplies	.03	.03	.03	.03	.08	.17
Production wells	1.12	1.12	1.12	1.12	2.80	6.31
Sulphur loading	.57	.57	.57	.57	.57	.57
State severance tax	1.03	1.03	1.03	1.03	1.03	1.03
Total	3.84	3.85	3.86	3.85	7.24	14.28
Fixed costs:						
Maintenance	0.12	0.16	0.21	0.16	0.40	0.91
Operating labor	.21	.30	1.20	.30	.75	1.60
Supervision	.04	.08	.32	.08	.20	.45
Labor and plant burden	.25	.38	1.52	.38	.95	2.14
Prop. taxes, insurance	.07	.09	.12	.09	.24	.53
Depreciation	.27	.36	.46	.36	.90	2.02
Total	0.97	1.38	3.83	1.38	3.44	7.74
Total unit costs	4.81	5.23	7.69	5.23	10.68	22.02

NOTE: The above are engineering cost estimates, made by the author; they are not intended to represent unit costs for any existing Frasch sul- phur plant. It was assumed in making these estimates that each plant was operating at full capacity.

ment with published estimates of Frasch production costs. For example, Jefferson Lake Sulphur Company has reported the following costs for its Long Point Dome plant with a capacity of 3.5 million gallons per day:[6]

	Output	Operating costs	Water ratio
	long tons	*$/long ton*	*gal./long ton*
1961	230,042	13.63	4,905
1962	255,964	13.21	4,549
1963	234,362	12.74	4,909

There operating costs include royalties, severance taxes, and all other taxes except income. Estimated royalty payments on sulphur produced at Long Point during these years were equivalent to $3.81, $4.10, and $2.78 per ton.[7] When these payments are deducted, the resulting operating costs ($9.82, $9.11, and 9.96) appear consistent with the slightly higher estimate of $10.68 given in table 7 for a plant of similar size and a dome of comparable quality (case II—plant 2).

From table 7, case I, it can be seen that plants with identical water ratios experience decreasing average costs with increases in plant size. The reduction is due almost entirely to the decrease in average fixed costs that accompanies increases in plant scale. Only very minor savings in variable costs can be attributed to larger-scale plants. Case II demonstrates that average costs for plants of the same size operating at full capacity vary directly with the water ratio experienced in mining. The reason for this is that total costs, with the exception of sulphur loading charges and state severance taxes, are identical for each plant, while the amount of sulphur produced over which these costs can be spread varies inversely with the water ratio.

Thus far, the analysis of unit costs has been limited to full capacity operation. Figure 4 presents the average total unit costs for the three plants described above for operations ranging from 50 per cent to 100 per cent of capacity. The dashed line in figure 4 connects the empirical minimum average total unit cost points on the cost curves of the successively larger Frasch plants (which are based on 100 per cent capacity operations), and conveys some idea as to cost variation related to size.

6. This information was disclosed at the time of the acquisition of Jefferson Lake by Occidental Petroleum. See Occidental Petroleum Corporation, *Listing Application A-21482*, New York Stock Exchange, 23 January 1964, p. 38.

7. Jefferson Lake leases the Long Point Dome from Texas Gulf Sulphur under a contract which, as amended 1 October 1960, calls for royalty payments of the greater of: (a) 50 per cent of the annual net profits from the mine, or (b) $2 per long ton. The royalty estimates given above were obtained by dividing the payments to Texas Gulf by the annual output.

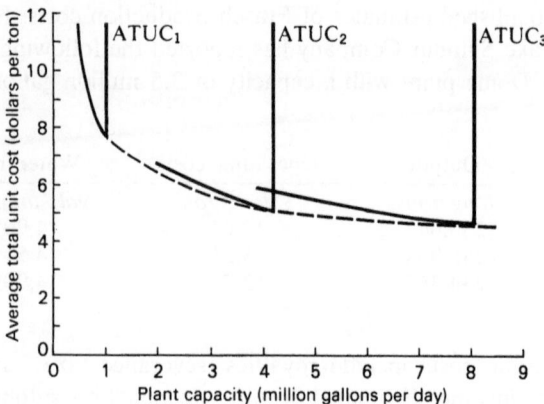

Figure 4. Effect of Frasch plant size and rate of operations on average total unit costs.

While unit costs continue to decline up to the maximum-size plant considered in this analysis, the flatness of the curve suggests that average total unit costs fall very slowly as plant size is increased beyond eight million gallons per day, and that the most significant economies are realized by the time the capacities in the range of four million gallons per day are reached.

In summary, this analysis of Frasch sulphur unit production costs has revealed the existence of substantial internal economies of scale. These economies result almost entirely from the savings realized in the use of fixed factors as plant size is increased and exhibit the common characteristic of increasing at a diminishing rate. They are, however, greatly overshadowed by the effect of ore deposit quality. The decline in unit costs accompanying declines in the water ratio (which reflect ore quality) over the actual range of water ratios experienced by firms in the industry is much more pronounced than the decline in unit costs accompanying increases in plant size, also taken over the range of existing plants. Finally, the high proportion of variable costs indicates that the penalty paid for operation at less than full capacity is not severe, at least not for plants of larger size.

Conditions of Entry into the Sulphur Industry

The sulphur industry has been characterized by extreme concentration throughout its history. In the United States, only five firms have achieved any significant production of Frasch sulphur, and two of these firms have produced over 85 per cent of the total Frasch sulphur output in this century. At the same time, several unsuccessful attempts have been made to enter the industry. The existence of only a few firms in an industry over a considerable time period leads one to inquire why other firms have not entered the industry. In a competitive situation, entry of new firms occurs in response to the long-run rate of profits; if the long-run rate of profit earned by existing firms is in excess of what can be earned in other industries in the economy, new firms should be attracted into the industry. If new firms have not entered an industry over time, analysis should then reveal that either (1) the long-run rate of profit earned by established firms in the industry has not been excessive; or (2) the established firms in the industry have significant advantages over potential entrant firms, these advantages being reflected in the degree to which the established firms can consistently maintain their prices above a competitive level without attracting new firms to enter the industry.

Profits earned by Frasch sulphur producers are analyzed in chapter 7. It is essential, however, before concluding the analysis of the market structure of the Frasch sulphur industry to determine if substantial advantages exist for established firms which would enable them to maintain a price level in the long run which would yield excess profits. These advantages, commonly called barriers to entry, have been classified into three groups: absolute cost barriers, product differentiation, and scale economies.[8] The existence of these barriers to entry is examined below.

Absolute cost barriers

If a potential entrant would have higher long-run costs than its established counterparts for any common scale of operations, the established firms can be said to possess an absolute cost advantage, and entry into the industry may be substantially restricted.

One source of such a cost advantage in the sulphur industry is the control by established firms of Frasch sulphur deposits. Since the sulphur industry is essentially a mining venture, holding a deposit is a necessary prerequisite to entry. However, Frasch sulphur deposits have been few in number and tightly controlled.[9]

The major Frasch sulphur producers acquired the sulphur rights to salt domes from the parties that had explored them for oil. Investigation of the relationship between the two largest sulphur producers and the major Gulf Coast oil companies reveals close ties between Freeport Sulphur and Texaco (formerly The Texas Company) and between Texas Gulf Sulphur and the Gulf Oil Corporation.

8. This classification follows Joe S. Bain, *Barriers to New Competition* (Cambridge: Harvard University Press, 1956), p. 12.

9. Strictly speaking, the nonavailability of Frasch sulphur deposits is an "absolute" barrier to entry only in the sense that the deposits which are not under the ownership of existing producers are of marginal quality, characterized by high water ratios, and having substantially higher unit costs than domes of better quality, even given the same scale of operation.

Freeport acquired at least four of the nine deposits it has mined from the Texas Company. In addition, the two companies have reciprocated directors. Since 1922 when Freeport leased the Hoskins deposit from the Texas Company with royalty payments of 70 per cent of the net profits from the mine, either the president or the chairman of the Texas Company has been a member of Freeport's board of directors. Since 1952, with the exception of 1956, two of Freeport's directors have also served on Texaco's board, and the chairman of Freeport has been a director of Texaco since 1955.

Texas Gulf Sulphur acquired at least three of the six deposits it has mined in the United States from the Gulf Producing Company, a subsidiary of Gulf Oil Corporation. From 1934 to 1948, Gulf Oil owned about 39 per cent of the outstanding Texas Gulf Sulphur stock. It continued to hold as much as 10 per cent until 1957. More recently, the two firms engaged in a joint venture to conduct offshore sulphur exploration.

Both Freeport and Texas Gulf, though closely associated with Texaco and Gulf Oil respectively, have also acquired deposits from Humble Oil and Refining and other Gulf Coast oil firms. As far as can be determined, no major petroleum company now owns as much as 10 per cent of the outstanding stock of either major sulphur producer.

Unlike their competitors, the two smaller Frasch sulphur producers were originally engaged in petroleum exploration. When their explorations turned up sulphur rather than oil, they entered the industry as sulphur producers. Both firms, however, were able to continue in the industry after their initial mines were exhausted by obtaining leases on marginal sulphur deposits from Texas Gulf Sulphur.

There are several reasons why more petroleum companies did not enter the industry. First, during the period when most of the salt domes were being discovered, petroleum exploration offered a high return on invested capital. The early wildcat drillers of the Gulf Coast were risk takers, and the promise of an above-average rate of return from sulphur production could not compete with the lure of the enormous potential return from a major oil strike.

Second, the Frasch sulphur companies were willing to pay a high price for the sulphur rights on salt domes with potential. Most contracts called for a royalty payment of 50 per cent of the net profits after the development expenses had been recovered. Thus, petroleum companies, without investing additional exploration and development funds, could reap 50 per cent of the benefits from sulphur production on their lands.

Third, after two firms were established in the industry with technical knowledge of the tricky Frasch process, access to markets, and, most importantly, tremendous aboveground inventories, successful entry became more difficult. This situation not only deterred entry by the oil companies but encouraged them to lease to the established firms rather than to potential entrants. Furthermore, the petroleum companies had no need to integrate backward into sulphur, as it was not until their acquisitions of agricultural chemical companies in the late fifties and early sixties that they became major consumers of sulphur.[10]

Product differentiation barriers

In some industries the preference of buyers for some or all of the products of established firms gives those firms a "product differentiation" advantage. However, as sulphur is a relatively homogeneous commodity purchased by professional buyers for large industrial firms, there is little opportunity for producers to differentiate their product. Thus, product differentiation has not acted as a barrier to entry in the sulphur industry.

Economies of scale barriers

The final barrier to entry, and the one that has perhaps received the greatest amount of attention in industry studies, is the barrier of economies of scale. If there are significant economies of scale throughout the range of possible industry output, the number of firms able to exist in the industry may be severely restricted. "Significant economies" means situations where the minimum optimal scale for a plant or firm in a particular industry, i.e., the smallest scale at which a plant or firm may achieve the lowest attainable unit cost, is a significant fraction of the total scale or capacity of the industry, and where unit costs at smaller than minimum optimal scales are significantly higher. In such cases, the entry of an additional firm may induce established firms to lower their price in order to preserve their market share. The entrant firm then faces the choice of entering the industry at a scale below its minimum optimal scale or entering the industry at its minimum optimal scale and facing a price after entry which is below the price prevailing prior to entry.[11] Any discussion of scale economies must be related to the plant

10. Thomas O'Hanlan, "The Great Sulphur Rush," *Fortune* (March 1968), p. 111, states: "Now, of course, with oil companies short of sulphur for their chemical and fertilizer divisions, the comfortable old camaraderie is over. From here on it's finders keepers."

11. There are actually two effects of scale economies on the conditions of entry. The one discussed here is known as the *percentage effect* and reflects the importance of the proportion of industry output supplied by an optimal sized plant or firm. A second effect, known as the *absolute-capital-requirements effect*, reflects the impact of large absolute amounts of capital investment required for optimum operation on the ease of entry into the industry. Since the absolute amount of capital required for construction of an optimal size Frasch sulphur plant is small, this effect has been omitted from the discussion of economies of scale.

and the firm, which are the functional units to which the economies are applicable.

The analysis of unit costs in the Frasch sulphur industry revealed economies of scale on the plant level, particularly up to the range of four million gallons of water per day. However, these economies do not appear to be large enough, relative to the demand for sulphur, to prevent the existence of a considerable number of plants in the industry. Annual domestic sales of Frasch sulphur in recent years have averaged over five million tons. If it is assumed that a plant with a water capacity of four million gallons per day can produce 500,000 tons of sulphur annually (reflecting an average water ratio of about 3,000 gallons per ton), then it would appear that a sizable number of "optimal size" plants could survive. Furthermore, the existence of several plants below this optimal size suggests that economies of scale at the plant level are not so large as to preclude profitable operation of suboptimal sized plants.

However, though about ten mines have generally been producing Frasch sulphur in the postwar period, these mines have been controlled by only four firms. Is it possible that larger economies of scale exist at the firm level? Certainly, multiplant firms have some significant advantages in the Frasch sulphur industry. The geographic centralization of Frasch mines makes it possible for a firm to construct a central terminal to serve several mines. Both Texas Gulf and Freeport conduct such an operation. It has been estimated that the installation of central handling and shipping facilities by Texas Gulf at its Beaumont terminal resulted in a 30–40 per cent reduction in its handling and shipping costs.[12] In addition, it is reasonable to assume that certain functions of the firm, such as research, financing, legal work, advertising, accounting, do not increase proportionally as the scale of the firm increases. It is therefore reasonable to assume that some economies of scale exist in these nonproduction functions of the firm. However, no evidence is available to suggest the magnitude of these firm economies. As has been the case in other industries, judgment as to the importance of economies of scale at the firm level must rest on indirect evidence. The existence of two small producers for over a twenty-five year period would seem to imply that economies of scale at the firm level are not of crucial importance in the Frasch sulphur industry, or that they are more than balanced by other factors, not necessarily connected with technology or economies, that elude analysis. However, survival is an imperfect test of efficiency because other elements, including antitrust legislation, can be respon-

12. See "Liquid Sulphur Terminal for TGS Cuts Costs, Eases Handling," *Sulphur* (February 1961), p. 27.

sible for the continuing existence of small firms in a concentrated industry.

General Supply Conditions for the Frasch Sulphur Industry

Sulphur production using the Frasch process is basically a mining operation. As such, its limits in time are set by the exhaustion of the ore deposit which is a nonreproducing form of capital. Nor (as Hotelling and Herfindahl have shown)[13] are these circumstances altered by long-run systematic exploration. However, the actual supply conditions of the Frasch sulphur industry must still be examined. In a perfectly competitive industry, the supply conditions could be shown by means of an industry supply curve depicting the lowest cost at which any given output could be produced. In the short run, the industry supply curve is simply the horizontal summation of the marginal cost curves of all firms in the industry. The long-run supply curve of the competitive industry will have a shape and position determined by the slopes of the marginal cost curves of the individual firms and the price at which each firm enters or leaves the industry.

However, Frasch sulphur is produced in an industry of few sellers. In such a situation, the industry supply curve is determined not only by the marginal cost curves of the individual firms, but also by the effect of a given shift in demand on the shape, slope, and position of the demand for individual producers. It is possible to isolate the supply response of the industry to a given change in price by constructing a schedule of what output would be forthcoming if the industry was given a set of alternative prices by an external source. For example, if the government agreed to purchase the entire output of the industry at various prices, the amount of output forthcoming at each price could be calculated by summing horizontally the marginal cost curves for all firms in the industry. In this manner, a supply price schedule could be derived. The curve reflecting such a schedule for the Frasch sulphur industry is given in figure 5. *ABC* is the short-run supply price schedule drawn to exclude the possibility of expanding the capacity of existing plants, but to include the possibilities of increasing production by substituting variable factors for the fixed capital, which is formed both by the ore deposit and by the plant associated with it.

The nearly flat portion of the curve, *AB*, represents production from the large, low-cost onshore deposits such as Boling and Grande Ecaille.

13. See Harold Hotelling, "The Economics of Exhaustible Resources," *Journal of Political Economy*, vol. 39 (April 1931), pp. 137–75; and Orris C. Herfindahl, "The Long-Run Cost of Minerals," *Three Studies in Minerals Economics* (Washington, D.C.: Resources for the Future, Inc., 1961).

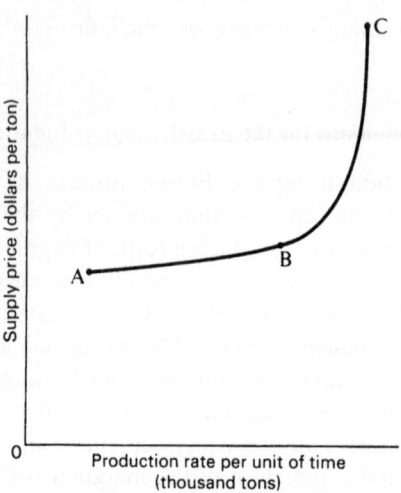

Figure 5.   Supply price schedule for Frasch sulphur.

The rising portion, *BC,* depicts the production of increasingly higher-cost sulphur from poorer quality mines. Between 1963 and 1965, in the face of rising demand and increased prices, the four major Frasch producers were able to expand output by about 25 per cent. However, without opening new mines or increasing the capacity of existing plants, it is unlikely that the four firms could increase their total output by more than 10 per cent. Thus, the supply of Frasch sulphur in the short run appears relatively inelastic. Additional output of Frasch sulphur in the short run will be accompanied by significant increases in cost.

Of course, it does not follow that, because the Frasch sulphur industry is one of increasing cost in the short run, the supply price for Frasch sulphur must rise over time. It is reasonable, however, to expect that Frasch sulphur production is subject to diminishing returns in the long run as well. Increasing costs in the long run would result from resource depletion, which causes upward shifts in the industry's aggregate marginal cost curve. In the case of Frasch sulphur, the easily accessible sources of supply have already been explored. Barring the discovery of new onshore salt dome formations, it would appear that increments to sulphur reserves must come from development of offshore salt dome deposits or from working the known onshore deposits that heretofore have been regarded as submarginal. In either instance, additional reserves could be developed only at an increasing cost. Of course, technological advances in exploration or mining could offset the effects of depletion as has happened in other minerals, such as copper, for example. It must be stressed that the above discussion pertains only to the long-run supply of *Frasch* sulphur. The case is much less certain for the long-run supply for all forms of sulphur. While domestic Frasch sulphur reserves appear to be diminishing, recent discoveries of sour natural gas in Canada, France, and the Middle East, and Frasch sulphur deposits in Mexico, coupled with continuing advancement in sulphur recovery techniques promise adequate sulphur supplies in the foreseeable future. The effect of these developments on the long-run supply price is discussed in chapter 8.

Chapter 4

Market Structure of the Sulphur Industry: Demand Aspects

This chapter is concerned with an analysis of the structural elements of the demand for sulphur that significantly affect the conduct of Frasch sulphur firms. These elements include (1) the number and type of consumers; (2) the distribution system and marketing procedures of Frasch producers; and (3) the cyclical and secular nature of the demand for sulphur. Together these factors define the demand environment within which firms in the industry must operate.

The Number and Type of Consumers

With the exception of its direct application as a plant nutrient, sulphur enters into various processing and manufacturing industries as a raw material or factor of production. Over 85 per cent of the brimstone consumed in the United States is burned to form sulphur dioxide for conversion to sulphuric acid. Thus, the demand for sulphur is largely dependent upon, or derived from, the demand for the end products of industries that consume sulphur either directly or in the form of sulphuric acid. A partial listing of industries and products in which sulphur is employed is given in table 8.

The sulphuric acid industry

The percentage of sulphur going into the manufacture of sulphuric acid has increased in recent years, rising from 79 per cent in 1950 to 87 per cent in 1966. Sulphuric acid may be manufactured from elemental sulphur, pyrites, and smelter gases, or from hydrogen sulphide derived from the purification of coke-oven and refinery gases; it may also be refortified or obtained by burning spent acid in sludge conversion plants to form new acid (see table 9).

When by-product sulphur is unavailable, the acid producer must choose between using elemental sulphur or pyrites as his raw material input. His choice is not determined completely by the relative prices of

Table 8. Industries and Products in Which Sulphur Is Used

Acids	Inorganic or organic acids
Alcohols	Insecticides
Alum	Leather
Ammonium sulphate	Livestock food
Aniline	Lubricants
Bleaching agent	Magnesium
Bromine	Matches
Carbon dioxide	Medicine
Carbon disulphide	Metallurgy
Carbon tetrachloride	Paints and pigments
Casein	Paper pulp
Cellophane	Petroleum products
Celluloid	Pharmaceuticals
Cellulose esters	Phenol
Cements	Photography
Chlorine	Plastics
Coke	Plate glass
Copper	Rayon
Dehydrating agent	Refrigerants
Detergents	Resins
Dyes	Road-surfacing materials
Ebonite	Rubber goods
Electroplating	Soap
Explosives	Soda
Fertilizers	Solvents
Fire extinguishers	Steel pickling and galvanizing
Fireproofing agents	Storage batteries
Fireworks	Sugar
Food preservatives	Sulphonated oils
Fumigants	Synthetic fibers
Fungicides	Synthetic rubber
Glue	Textiles
Glycerin	Tires, rubber
Impregnant	Water purification

SOURCE: U.S. Bureau of Mines, *Mineral Facts and Problems, 1965 Edition*, Bulletin 630, p. 908.

Table 9. Domestic Sulphuric Acid Production by Raw Material Source

per cent of total acid

Year	Elemental sulphur	Pyrites	Smelter gases	Hydrogen sulphide	Reforti-fied	Acid sludge
1920	51.76	22.92	25.32			
1930	58.48	22.05	19.47			
1940	62.12	25.57	11.44	0.87		
1950	73.75	11.51	5.69	1.00	6.81	1.24
1960	69.84	10.46	6.62	1.84	4.46	6.78
1966	75.58	4.92	5.73	3.42	3.41	6.94

SOURCE: *Chemical Economics Handbook* (Menlo Park, California: Stanford Research Institute, 1961 and 1967), No. 792.5031 and No. 792.2010A–8030G.

these two raw materials at their source. Transportation costs are also important because only about 45 per cent of the weight of pyrites usually becomes available for acid production, while all of the weight of elemental sulphur is available. Another consideration is that acid production from pyrites results in a by-product sinter, having iron and other metal values. But freight costs are a factor here too, as the sinter must be moved to an appropriate smelter.

Because of its purity, elemental sulphur can be converted to sulphuric acid in a relatively simple and inexpensive acid plant. However, an acid plant designed to use pyrites is more complex and requires 30 to 60 per cent more investment than a comparable plant built to consume elemental sulphur,[1] and total factory costs are also greater.[2]

Sulphuric acid plants are widely scattered throughout the United States, chiefly because of the low bulk value of the acid, the difficulties of handling the acid in bulk, and the subsequent high costs of shipment. These factors have also contributed to the conspicuous absence historically of vertical integration in the sulphur industry. Few significant economies can be obtained from centralizing the production of sulphuric acid at or near Frasch sulphur mines.

Most acid produced is not sold on the open market. Large users, such as fertilizer, chemical, and explosives manufacturers, usually own and operate their own acid plants in locations determined by proximity to major markets or to other raw materials. In 1965, it is reported that 94 companies were producing sulphuric acid at 230 plants located in 42 states and Puerto Rico.[3] Although many of these companies are industrial giants, no single acid producer uses enough sulphur to exercise significant control over the sulphur market.

End-use analysis of sulphur consumption

The relative importance of the various end uses for sulphur in acid and non-acid forms is indicated in table 10 where domestic sulphur consumption is broken down by major industry. Four comments are in order regarding the nature of this breakdown of sulphur consumption by industry.

First, only a small proportion of sulphur is consumed directly in ground and refined form as a secondary plant nutrient, a fungicide, and

1. W. L. Faith, Donald B. Keyes, and Donald L. Clark, *Industrial Chemicals* (3d. ed.; New York: John Wiley and Sons, Inc., 1965), p. 753.
2. See "Rethinking Rejected Routes," *Chemical Week* (14 October 1967), p. 122.
3. Faith, Keyes, and Clark, *op. cit.*, pp. 754–56. Because a number of companies produce acid for use within their own plants, it is difficult to obtain accurate figures on the exact number of acid producers. Thus, these figures are only an approximation.

an insecticide; the rest is consumed indirectly in the production of other commodities. The demand curve of a firm for sulphur, as an intermediate product, depends upon four factors: (1) the technology, i.e., both the marginal rate of substitution between sulphur and alternative factors, and the relationship between marginal physical product and changes in sulphur input; (2) the nature of demand for the firm's products; (3) the prices of other factors; and (4) the elasticity of supply of other factors employed. While it is difficult to analyze each of these factors for the many firms using sulphur in one form or another, it is possible to make some broad judgments on the basis of the industries shown in table 10.

Table 10.   Distribution of Domestic Sulphur Consumption, 1966

Consuming sector	Per cent of total sulphur consumption
Acid uses:	
Fertilizers	48
Chemicals	18
TiO$_2$ and other inorganic pigments	6
Iron and steel	3
Rayon and film	3
Petroleum	2
Others	7
Total acid uses	87
Nonacid uses:	
Pulp and paper	5
Carbon bisulphide	3
Ground and refined	2
Other	3
Total nonacid uses	13

SOURCE: Freeport Sulphur Company, *Sulphur—Ally of Agriculture and Industry* (1967).

The marginal rate of substitution between sulphur and other factors is relatively high in such industries as chemical fertilizers, paint and pigments, petroleum, iron and steel, and pulp and paper, where technically and economically feasible substitutes exist. On the other hand, the marginal rate of substitution is relatively low in the nonferrous metals, rayon, and rubber industries, where substitutes are technically inferior to sulphur or sulphuric acid. Thus, it would be expected that the elasticity of demand for sulphur would be higher for firms in the first group of industries than for firms in the second group, given the time required for substitution to occur.

In most cases, sulphur as sulphuric acid is combined in almost fixed proportions with other inputs; thus, the marginal physical product declines rapidly as additional units of sulphur input are added. For this reason, small price declines for sulphur will in general fail to increase sulphur consumption, particularly in the short run where only the input of sulphur can be altered with other factor inputs remaining fixed. This would establish a presumption in favor of an inelastic demand for sulphur in the short run.

The second comment regarding the breakdown of sulphur consumption given in table 10 is that in most of its uses sulphur represents only a small fraction of the total cost of the products in which it is used. For example, 100 pounds of sulphur (costing about $1.80 in 1967) are used in the production of 800–1,000 pounds of black powder, 2–2.5 tons of nitroglycerine, 100–125 pounds of viscose rayon, 5,000 pounds of soft rubber, and 885–1,060 pounds of superphosphate fertilizer.[4] As Stigler has pointed out, it cannot be said in general that the elasticity of demand for a factor of production is smaller the smaller the proportion of total cost the outlay on the factor is, without adding the qualification, "given the possibility of substituting other inputs."[5] For the sulphur industry, the statement with qualification added appears valid, particularly in the short run. A small or even moderate change in the price of sulphur is unlikely to induce short-term adjustments in the amount of sulphur used because a shift to a different input would involve changes in plants, processes, products, input proportions, and expenditures that would generally be more costly than any savings likely to result from substitution. Thus, the fact that sulphur in general accounts for only a small portion of the total cost of most of the products in which it is used, coupled with the unlikelihood of substitution in the short run, supports the presumption of an inelastic demand for sulphur in the short run.

In the long run, however, the demand for sulphur is probably elastic at least over some range of prices. The continued use of sulphur as an industrial raw material is dependent to a large extent on how the price of sulphur and the costs of sulphur-consuming processes compare with the price of technical substitutes and the costs of processes employing these substitutes. Unique uses for sulphur are few. In most instances, technical substitutes are available. In addition, since there are several

4. Texas Gulf Sulphur Company, *Facts About Sulphur* (New York, 1958) p. 24.

5. George J. Stigler, *The Theory of Price* (rev. ed.; New York: The Macmillan Company, 1952), pp. 190–91. Stigler asserts that "the smaller the fraction of total cost that a productive service constitutes, the more finely the productive service is classified as a rule, and the more finely the productive services are classified, the easier it is to substitute other productive services." Thus, he feels that the qualification is important, and "robs the statement of much of its importance."

sources of sulphur, the demand for any single source of sulphur (for example, Frasch sulphur) is more elastic than the demand for sulphur in general. Thus, the possibility of substitution tends to make the long-run demand for sulphur relatively price elastic.

However, it must be remembered that a decision to use sulphur generally coincides with a decision to use a particular process. Thus, the purchase decision has investment overtones. Changing from elemental sulphur to a technical substitute, or to another form of sulphur, generally will involve a change of processes, requiring both changes in input proportions of other raw materials and changes in plant and equipment. As price rigidity reduces the element of risk in a purchase decision with investment implications, purchasers of sulphur are felt to prefer a reliable source of sulphur at a reasonably stable price to widely fluctuating prices.[6]

The final comment regarding the end uses of sulphur by consuming industries is prompted by the observation that sulphur is used in a wide range of industries covering most segments of American industry. Liebig's axiom that the economic welfare of a nation can be most accurately measured by its consumption of sulphuric acid appears to be as valid today as it was when he advanced it in the middle of the nineteenth century, though we tend to use the term "welfare" more cautiously today. The diversity of uses for sulphur as an intermediate product subjects the demand for sulphur to the cyclical fluctuations that characterize the manufacturing segment of the economy. At the same time, it makes the long-run demand for sulphur dependent upon growth in industrial production. While sulphur is sold in relatively small lots to a great many firms, ranging in size from small independent acid manufacturers to large integrated concerns, one major Frasch producer estimates that its largest ten to fifteen customers account for 90 per cent of its domestic sales. Nevertheless, it is unlikely that any single purchaser constitutes enough of the sulphur market to be able to wield significant power over that market.

Export markets for sulphur

Sulphur consumption provides a reasonably good index of a country's stage of industrialization. For example, per capita consumption of sulphur in 1965 amounted to 95 pounds in the United States, as compared with 7 pounds in Latin America and only 2 pounds in India.[7] The dis-

---

6. See *Sulphur* (September 1957), p. 16. Peck and Phillips in their respective studies of the aluminum and synthetic rubber industries observe that price rigidity is often desired by consumers of intermediate products.

7. Estimates made by Freeport Sulphur Company.

tribution of free world consumption of sulphur in 1968 by geographic area and sulphur source is shown in table 11. As in the United States, the demand for sulphur worldwide is closely related to the production of sulphuric acid.

Table 11. Free World Consumption of Sulphur, 1968, by Geographic Area and Sulphur Source

thousand long tons, sulphur equivalent

Area	Elemental	Nonelemental	Total
North America			
United States	8,275	1,025	9,300
Canada	725	550	1,225
Subtotal	9,000	1,575	10,525
Latin America	725	75	800
Europe	3,700	5,550	9,250
Africa and Middle East	550	450	1,000
Far East	850	2,400	3,250
Oceania	650	275	925
Total, free world	15,475	10,275	25,750

SOURCE: *Engineering and Mining Journal* (March 1969), p. 160c.

The major export markets for Frasch sulphur traditionally were Canada, the United Kingdom, France, Brazil, India, and Australia. The advent of the recovered sulphur production in Canada and France and Frasch sulphur production in Mexico brought changes in the pattern of world trade in sulphur (see table 12). The U.S. share of total world sulphur trade fell from about 47 per cent in 1960 to less than 30 per cent in 1964. During the same period, Mexican Frasch producers managed to maintain a nearly constant share of about 30 per cent, while French and Canadian recovered sulphur producers significantly increased their share of world sulphur markets. In 1965, governmental export restrictions and production problems caused a sharp reduction in exports of Frasch sulphur from Mexico, and U.S. Frasch producers were able to reverse the downward trend in their export share.

From 1922 to 1952, most export sales of Frasch sulphur were made through the Sulphur Export Corporation (Sulexco), a Webb-Pomerene export association. Sulexco was disbanded in 1952, largely because a worldwide shortage of sulphur made such a sales organization superfluous. In 1958, in the face of growing competition in many of their traditional export markets, the four domestic Frasch producers reestablished Sulexco.

International trade in sulphur has for a long time been subject to

Table 12. World Sulphur Trade, 1960–65

thousand metric tons

Source and destination	1960	1961	1962	1963	1964	1965
U.S. Frasch (% of export market)	46.7%	41.4%	36.9%	30.9%	29.5%	35.5%
North America	301	307	180	144	138	153
Western Europe	733	632	626	626	712	1,214
Australia	238	216	199	229	228	276
S. and Cent. America and Caribbean	204	210	198	195	313	232
Asia	199	170	262	312	367	393
Africa	93	52	47	33	11	99
Communist countries	34	22	45	53	50	35
Total	1,802	1,609	1,557	1,592	1,819	2,402
Mexican Frasch (% of export market)	30.8%	29.6%	31.4%	28.4%	30.5%	23.9%
North America	599	650	735	870	917	789
Western Europe	293	249	308	353	667	527
Australia	122	105	95	135	71	143
S. and Cent. America and Caribbean	15	4	22	21	19	18
Asia	68	56	61	13	28	22
Africa	88	89	103	76	131	120
Communist countries	4	—	—	—	51	—
Total	1,190	1,153	1,324	1,468	1,884	1,619
France (Lacq) (% of export market)	10.8%	15.0%	12.5%	19.6%	15.7%	14.0%
North America	—	—	—	12	—	—
Western Europe	349	441	433	700	835	796
Australia	—	5	5	12	—	—
S. and Cent. America and Caribbean	8	6	6	49	9	—
Asia	10	45	13	21	8	8
Africa	44	59	48	129	92	125
Communist countries	6	25	24	87	23	17
Total	418	582	529	1,010	967	947
Western Canada (% of export market)	3.4%	5.0%	8.6%	14.4%	19.0%	20.0%
North America	130	188	297	485	574	681
Western Europe	—	—	15	17	103	66
Australia	—	—	22	51	174	214
S. and Cent. America and Caribbean	—	—	—	3	33	38
Asia	—	6	29	106	136	162
Africa	—	—	—	29	48	91
Communist countries	—	—	—	54	106	105
Total	130	194	363	745	1,174	1,357
Other exporters—free world	114	80	66	37	91	34
Communist bloc	208	275	388	312	240	422
World total	3,862	3,893	4,222	5,170	6,177	6,781

SOURCE: *Sulphur*, various issues.

NOTE: Totals may not add due to rounding.

controls and restrictions. For example, imports into the United Kingdom, Australia, and New Zealand are handled by purchasing cartels. In addition, some nations such as Japan and Italy traditionally have placed import quotas or high tariffs on sulphur in order to protect their domestic high-cost native sulphur ore and pyrites industry. While these restrictions have been relaxed somewhat in recent years, they still reduce sulphur sales in some export markets. Mexico, however, has raised tariff barriers in recent years, following the development of its domestic Frasch industry in the 1950s.

Distribution and Marketing of Frasch Sulphur

Markets, like people, appear to be creatures of habit. Forms of distribution and marketing practices, once established, usually persist for considerable periods of time. They play important roles in defining the nature of the price-making process and in determining the character of the responses by firms to changes in supply and demand. The absence of an organized commodity exchange for sulphur makes even more necessary a consideration of the distribution system and marketing practices of Frasch firms. In particular, it raises questions concerning the availability and validity of pricing data, the dimensions of price, and the degree of information concerning market conditions possessed by both sellers and buyers.

Institutional aspects of the Frasch sulphur market

The most striking feature of the manner in which Frasch sulphur is sold to domestic consumers is the directness and simplicity of the system. While small amounts of sulphur bypass the market through transfers to other parts of integrated companies, most Frasch sulphur is sold by the producer directly to the consumer, and not through agents, brokers, or other middlemen. Within the United States, Frasch sulphur is marketed through district sales offices located in the major market areas (Tampa, Florida, the Atlantic Seaboard, the Great Lakes, the Gulf Coast, and the inland water routes of the Mississippi, Ohio, and Illinois rivers). As mentioned previously, all export sales of Frasch sulphur were made through the Sulphur Export Corporation until 1952, and again since 1955. Until Duval withdrew in 1963, the ownership of Sulexco was divided as follows: Texas Gulf Sulphur Company—37 per cent, Freeport Sulphur Company—37 per cent, Jefferson Lake Sulphur Company—18 per cent, and the Duval Corporation—8 per cent. Sulexco, in turn, maintains a worldwide sales organization and has agents in every major foreign sulphur market.

Sulphur is generally sold under contracts on both domestic and export markets.[8] The contracts usually run for one year or more, but are

---

8. Historically, contract periods have varied in response to expectations concerning future supply and demand conditions. When a short supply of sulphur has been anticipated, consumers have pressed for long-term contracts to insure availability of future deliveries. In the early sixties, when sulphur was abundant and prices weak, buyers attempted to shorten the contract period. Usually, sulphuric acid manufacturers desire a contract period of at least one year to insure themselves twelve months' supply. A factor causing longer contract periods in recent years has been the inclusion in sulphur contracts of provisions calling for conversion of the buyer's plant to the storage, handling, and use of molten sulphur, sometimes at the seller's expense. In these instances, contracts were written for a longer period, generally either five or ten years. Even in these cases, how-

not full requirement contracts. As a rule, they contain provisions granting the seller the right to increase the price by notification to the purchaser. The buyer then has a specified period, usually thirty days, in which to accept or reject the price increase. If the price increase is rejected, the seller has the right to cancel the contract. On the other hand, if the purchaser receives a bona fide offer from another seller at a lower price, he may cancel his contract after giving the original seller the option of meeting the lower offer. In addition, the seller has an obligation to extend to the buyer any lower price agreed upon by the seller for a comparable sale to another buyer in the vicinity of the initial buyer's plant.[9] Consumers usually call for bids and then negotiate a contract with the most favorable supplier. Contracts generally call for minimum quantities to be purchased and specify delivery dates. There is some evidence that most purchasers of sulphur refrain from obtaining all their sulphur requirements from one seller in order to enhance their bargaining position and to ensure a continuing source of supply.

Distribution of Frasch sulphur

The geographical concentration of Frasch sulphur production and the widely dispersed locations of its primary consumers make the distribution of sulphur from the mines to the major consuming areas an important feature of the industry's market structure. Until recently, sulphur was marketed primarily from the mines. Union Sulphur Company shipped its output by rail directly to consumers or to Sabine, Texas, where it was loaded on freighters for transshipment either to the Atlantic Seaboard or abroad. Later, Texas Gulf Sulphur, Freeport Sulphur, Jefferson Lake Sulphur, and Duval Sulphur and Potash also shipped much of their product directly to the consumer by rail, and their ocean shipments were loaded on freighters at Texas City, Galveston, and Freeport, Texas. In time, major distribution points for sulphur developed on the Atlantic Seaboard in the port cities of New York, Baltimore, Portland, Searsport (Maine), and Quebec. Similarly, on the Pacific Coast, Portland, Oregon, and Vancouver, B.C., became major sulphur distribution points. Abroad, sulphur was distributed from Marseilles, Manchester, Hamburg, and Gothenburg.[10]

---

ever, the buyer has the option of cancelling the contract under certain provisions, provided he pays the seller the unamortized balance of any conversion cost incurred by the seller.

9. This obligation is not usually contained in the written contract. However, in conversations with both buyers and sellers, the author has ascertained that such an obligation is felt to exist by both parties.

10. See Albert G. Wolf, "Marketing of Sulphur," *The Engineering and Mining Journal* (1 July 1922), pp. 19–20.

Traditionally, sulphur has been marketed to three major groups of consumers: the pulp and paper manufacturers of New England and Eastern Canada; the chemical manufacturers of the eastern states; and the acid phosphate manufacturers of the southeastern states. Other consumer groups have been the various industries of the Great Lakes region, and the explosives, pulp and paper, insecticide, and fungicide manufacturers of the Pacific Coast. This grouping of major consuming areas has not changed greatly since the early days of the Frasch sulphur industry. The only important change has been the growth in sulphur consumption in the central and southwestern states. This occurred when the chemical industry spread into these areas as petroleum and natural gas became increasingly important as a raw material source.

For many years, sulphur was shipped in dry bulk form to the consumer. From the mine, the sulphur was moved through steamheated pipes to central points called relay stations, where it was discharged into steamheated tanks and relieved of its contained air. From the relay stations, sulphur was pumped to the storage area where it was sprayed into vats to cool and solidify. When ready for shipment, the solid sulphur was broken up by power shovels, loaded onto conveyor belts, and moved to barges, freight cars, or ocean freighters for delivery to the consumer. At the consumer's plant, sulphur was again stored in dry bulk form until required for plant operations. In most instances, the sulphur had to be melted to liquid form before it could be used.

Beginning about 1955, some liquid shipments of sulphur were made by barge to consumers in Alabama, Arkansas, Mississippi, and Missouri via inland waterways. However, liquid shipments did not assume significance until after 1960, when major alterations were made in the distribution system for Frasch sulphur.

In 1958, Texas Gulf Sulphur Company constructed the first liquid storage terminal for sulphur at Cincinnati to serve customers in the Ohio Valley. In 1959, Texas Gulf began construction of a mammoth new $3 million storage and shipping terminal at Beaumont, Texas, on the Neches River. This terminal, completed in 1960, was designed to handle the bulk of the company's sulphur production from its four Frasch mines, both in the solid and liquid states. Rail, truck, barge, and ocean shipments could all be accommodated at this terminal. The centralization of shipping and storage facilities reduced Texas Gulf's handling and shipping costs by an estimated 30–40 per cent.[11] In addition, between 1958 and 1963, Texas Gulf constructed ten additional liquid storage terminals in major consuming areas. To carry liquid sulphur to the ter-

11. "Liquid Sulphur Terminal for Texas Gulf Sulphur Cuts Costs, Eases Handling," *Sulphur* (February 1961), p. 27.

minals in Tampa, Florida, and the Atlantic Seaboard, Texas Gulf chartered a converted T-2 15,000-ton tanker, the S.S. *Marine Sulphur Queen*, which went into service in January 1961. This was the first ocean-going vessel to be used exclusively for the transport of liquid sulphur.[12]

In mid-1959, the Freeport Sulphur Company launched a $23 million program to move sulphur in liquid form into major market areas for storage and transshipment to its customers. Freeport entered into long-term contracts with independent marine transportation companies to provide and operate new terminal facilities and marine equipment, estimated to cost about $20 million.[13] Additional facilities, costing between $3 million and $4 million were provided by Freeport. Freeport also leased two tankers, the S.S. *Louisiana Sulphur* and S.S. *Louisiana Brimstone,* to serve its domestic coastal markets.

In 1961, Pan American Sulphur Company constructed a liquid sulphur terminal at Tampa to handle part of its exports to the United States. PASCO later increased the size of the Tampa terminal and added a second terminal at Newark. In 1962, the other Mexican Frasch company, Gulf Sulphur Corporation, opened liquid sulphur terminals in Tampa and Baltimore.

In 1963, U.S. and Mexican Frasch producers shipped liquid sulphur to twenty-seven producer-controlled regional storage and transshipment terminals. (The location and the capacity of these terminals are listed in table 13.) The two small domestic Frasch firms, Jefferson Lake Sulphur and the Duval Corporation, ship liquid sulphur to customers in either tank trucks or railroad cars. Neither firm has built regional liquid sulphur terminals to handle water shipments. Duval ships only a small percentage of its sulphur in liquid form, and Jefferson Lake ships much of its liquid sulphur into areas where other producers also ship by rail or truck.

The movement of liquid sulphur to overseas markets began in 1964. Sulexco chartered two 25,000-ton tankers, the *Naess Texas* and the *Naess Louisiana,* and constructed an $18 million liquid terminal at Rotterdam, Netherlands, capable of handling approximately 500,000 tons of liquid shipments annually and storing more than 100,000 tons of solid sulphur. In 1965, Sulexco increased the capacity of its Rotter-

---

12. On February 2, 1963, the *Marine Sulphur Queen*, carrying a cargo of liquid sulphur, left Beaumont, Texas, for Norfolk, Virginia. Without distress signal, the ship disappeared with the tragic loss of its crew of thirty-nine. The disappearance remains a mystery to this day. In 1964, Texas Gulf chartered the 23,760-ton tanker, S.S. *Marine Texas* to replace the *Marine Sulphur Queen*.

13. See "Freeport Sulphur Company's Activities," *Sulphur* (February 1961), pp. 25–26; and "Trend to Molten Sulphur Grows Fast," *Chemical and Engineering News* (April 10, 1961), pp. 23–24.

Table 13. Liquid Sulphur Regional Storage and Transshipment Terminals in Operation in 1963

Producer-controlled terminals	Storage tanks	Total storage capacity
	number	*thousand long tons*
Freeport Sulphur Company:		
Baton Rouge, La.	1	6.5
Bucksport, Maine	2	20.0
Charleston, S.C.	1	10.0
Everett, Mass.	1	10.0
Joliet, Illinois	3	30.0
Nitro, W. Va.	2	18.0
Tampa, Florida	6	60.0
Warners, N.J.	2	12.5
Wellsville, Ohio	2	20.0
Total	20	187.0
Texas Gulf Sulphur Company:		
Baltimore, Md.	2	24.0
Carteret, N.J.	2	26.0
Cincinnati, Ohio	3	16.8
Jacksonville, Florida	1	11.0
Marseilles, Ill.	1	10.0
Newell, Pa.	1	10.0
Norfolk, Va.	2	20.8
Paulsboro, N.J.	2	24.0
Savannah, Ga.	1	11.0
Tampa, Florida	1	7.5
Wilmington, N.C.	1	8.0
Total	17	169.1
Pan American Sulphur Company:		
Newark, N.J.	1	10.0
Tampa, Florida	4	40.0
Total	5	50.0
Gulf Sulphur Corporation:		
Baltimore, Md.	1	10.0
Tampa, Florida	1	10.0
Total	2	20.0

SOURCE: U.S. Department of the Interior, Bureau of Mines, *Minerals Yearbook, 1965*, vol. 1 (Metals and Minerals—Except Fuels).

dam terminal by one-third and built a second liquid terminal at Dublin, Ireland. Customers are serviced from these terminals by self-propelled tanker barges and other carriers.

In 1965, PASCO also began liquid shipments abroad, placing in operation a 60,000-ton liquid sulphur terminal at Immingham, England. The company expects to ship about 225,000 tons of liquid sulphur annually through this terminal, which is serviced by three chartered tankers.

The change from solid to liquid shipment of sulphur occurred rapidly, with liquid shipments increasing from only 15 per cent of domestic shipments in 1959 to an estimated 90 per cent by 1963.[14] Producers had two major reasons for their sudden shift to liquid delivery. First, conversion to liquid sulphur made it possible to sell under longer-term contracts, which producers found desirable in a period of lagging sales and price competition. Conversion to liquid sulphur substantially reduced inventories held by both consumers and producers because of the added expense of maintaining the inventory in a liquid state. The reduction in producers' inventories made spot sales much more difficult. Distribution of sulphur from producer to consumer became subject to a greater degree of scheduling. As a result, producers were able to demand longer-term contracts. This trend to longer-term contracts was also encouraged by the willingness of sulphur producers to finance conversion of customers' plants to liquid sulphur in return for a longer-term purchase agreement. A second reason for the rapid switch to liquid delivery was that it required the use of special ships which reduce the freight advantage held by Mexican Frasch producers who were free to ship sulphur into the United States in foreign-registered tramp steamers.

Once the decision was made to move to liquid delivery of sulphur, it then became to the producers' advantage to convert as much of their shipments as possible to the new form of distribution. They were aided in this effort by the fact that liquid delivery benefited consumers also. Most sulphur, whether for acid or nonacid consumption, is used in liquid form, and liquid delivery eliminates fuel costs for sulphur melting and reduces handling costs. Liquid delivery also eliminates losses in handling, which average 0.5 per cent for dry sulphur. Furthermore, liquid sulphur is not subject to contamination from moisture, scale, and other foreign matter, and there is no dust to cause corrosion of adjacent equipment and create an explosion hazard.[15] It is estimated that

---

14. L. B. Gittinger, Jr., "Sulphur," *The Engineering and Mining Journal* (February 1964), p. 151.
15. John Doak, "Liquid-Sulphur Distribution," *The Oil and Gas Journal* (June 24, 1963), p. 101.

conversion of a plant to use of liquid sulphur lowers plant costs by as much as $2 per ton. In addition, conversion to liquid sulphur, combined with the establishment by producers of regional terminals, lowers the inventory requirements of sulphur consumers.

Nature of the price structure

As sulphur is normally sold under long-term contract with no formal market, both buyers and sellers must rely on their own initiative to collect price data. U.S. domestic list prices for Frasch sulphur are generally quoted per long ton, f.o.b., bulk, cars, mines; f.o.b., vessels, Gulf port (producer's port); and f.o.b., regional terminal. Prices are $1.50 lower at the mines than at the Gulf ports. The regional terminal prices vary with location. The current changes in freight prices do not affect the list prices at the mines and Gulf ports. U.S. export list prices, which are quoted per long ton, f.o.b., vessels, Gulf port, and c & f for certain destinations, are not based on domestic list prices, but are determined separately.

Two grades of Frasch sulphur are sold: bright (clear yellow sulphur not discolored by hydrocarbon impurities) and dark. Bright Frasch commands a $1.00 premium per long ton. Generally, the quoted list prices for spot and contract sales are the same. However, in the past, list price changes have been made effective on spot transactions and new contracts three to six months before they have been made effective for current contracts.

Fluctuations in the Demand for Sulphur

Seasonal fluctuations in the demand for sulphur do not exercise much influence on the behavior of firms within the industry. The demand for sulphur is, however, responsive to cyclical fluctuations in industrial production. This is illustrated in figure 6. The reason for this close relationship between sulphur demand and industrial production is the diversity of uses to which sulphur is put. Since sulphur is consumed in almost every sector of modern industry, the demand for sulphur tends to fluctuate with industrial activity in general, and "historical statistics show the remarkable harmony between the movement of sulphur consumption and that of industrial production."[16] It might be thought that the shifts in sulphur consumption that have occurred in recent years and, in particular, the increasing importance of the chemical fertilizer industry as the primary consumer of sulphur, would have weakened the

16. See Hans H. Landsberg, Leonard L. Fischman, and Joseph L. Fisher, *Resources in America's Future* (Baltimore: The Johns Hopkins Press for Resources for the Future, Inc., 1963), pp. 326–27.

Index (1957-59 = 100)

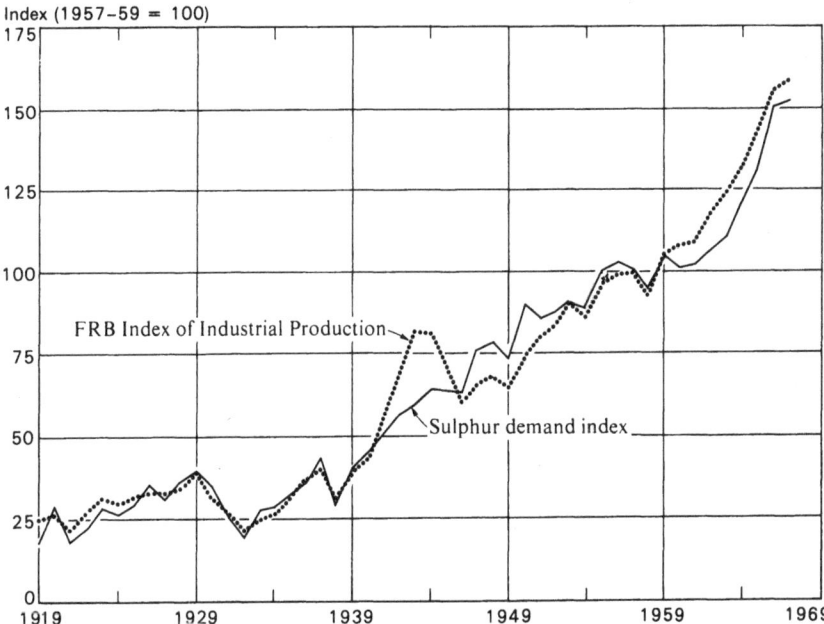

Figure 6. U.S. sulphur demand compared with industrial production, 1919–67.

historical relationship between sulphur consumption and industrial production. To test this hypothesis, two correlations of sulphur consumption with industrial production were made, the first for the period 1919–67 and the second for the period 1919–59. The resulting coefficients of determination were 0.9397 and 0.9675 respectively, indicating only a slight weakening in recent years of the relationship linking sulphur consumption and industrial production. Of course, it might be that the increased use of chemical fertilizers in the sixties has merely coincided with an increase in industrial production and that there is no permanent relationship between the two.

The demand for sulphur in the United States has increased significantly over time (see figure 7). The growth in consumption since 1952 shown in figure 7 is actually understated because it does not include (1) the sulphur values in suphuric acid reconstituted from acid sludges, or (2) reductions in consumers' inventories which accompanied the switch to liquid distribution of sulphur.[17] Exports, also shown in figure 7, have followed the general pattern of domestic consumption with the exception of the periods during World War II and Korea when exports

17. See William L. Swagger, *The Paley Report in Review: Sulphur* (Battelle Memorial Institute, 1961), pp. 5–6.

Figure 7.   U.S. apparent consumption and exports of sulphur, 1905–67. (Based on data in table A-2.)

fell and domestic consumption increased. The extent of the increase in exports in recent years may have been somewhat overstated by including shipments that were intended to build up producers' stocks at newly opened terminals in Europe.

Part II

Market Behavior in the Sulphur Industry

Part II

Market behavior in the Sapphire Industry

Chapter 5

Output and Prices

In the introductory chapter to this study, the thesis was presented that an industry's price and output behavior could be related to the underlying factors of the industry's market structure. If the market structure of the industry undergoes significant change, it should be possible to observe resulting changes in market behavior within the industry. Earlier chapters of this study showed that the market structure of the Frasch sulphur industry underwent significant alterations during the fifties. In particular, new sources of supply, both domestic and foreign, entered into competition with Frasch sulphur for domestic and export markets. Thus, the Frasch sulphur industry affords an opportunity to test the hypothesis linking market behavior to market structure. As the structural characteristics of the Frasch industry have changed, similar changes should have occurred in market behavior within the industry.

It should be stressed, however, that while the market structure of the Frasch sulphur industry has changed over time, conditions in the industry have always closely approximated those of pure oligopoly. There have never been more than six domestic firms producing Frasch sulphur at any given time, and four firms have virtually dominated the industry. Entry of new firms has been restricted by the availability of workable deposits. Frasch sulphur has remained an essentially homogeneous commodity. Thus, the structural conditions in the industry have resulted in a maximum degree of interdependence among firms in the industry. The major changes in market structure have related to the degree to which the actions of Frasch producers have been independent of external restraint.

Frasch sulphur must compete in the market with other sources of elemental sulphur (recovered sulphur and native sulphur) and with various forms of nonelemental sulphur. Over the life of the industry, the market actions of domestic Frasch producers have been influenced to varying degrees by actual and potential competition from these other sources of sulphur, both domestic and foreign.

This chapter seeks to explain the observed market behavior of Frasch producers by relating price and output decisions to the changing structural characteristics of the industry. Output behavior in the Frasch sulphur industry is considered in section one of this chapter and price behavior is examined in section two. The following chapter contains a chronological explanation of market behavior in the industry by linking changes in the form of oligopolistic market behavior to changes in the industry's market structure.

Output Behavior in the Frasch Sulphur Industry

Short-run aspects of Frasch sulphur production

While quarterly production figures for Frasch sulphur are available back to 1924, a monthly production series extends back only to 1943. In the following analysis, three series have been used for the period 1951 to 1968. The first series is the monthly production of Frasch sulphur. The second is the month-end producers' inventories of Frasch sulphur. The third, computed from the first two, shows monthly shipments.

The first step taken in analyzing these three time series was to determine the extent of seasonal variation. To test for seasonal fluctuations, a 12-month moving total of the monthly observations was computed. To obtain a monthly moving average, the preceding and succeeding moving totals were averaged. Specific seasonals for each month were then computed by taking the ratio of the monthly observation to the monthly moving average. These specific seasonals for each month were then averaged for the period 1951–68, and the resulting means were then "leveled" to insure that they summed to 1200. The resulting monthly index of seasonal variation is given in table 14.

Analysis of the monthly seasonal indexes reveals that inventories are subject to only very minor seasonal variation, while both production and shipments exhibit a somewhat greater degree of seasonal sensitivity. Since Frasch plants operate on a continuous 24-hour-day basis, some monthly variation results from the varying number of days in the month. Frasch plants operate 672 hours in February, but 744 hours in 31-day months. Additional random variation results from the opening and closing of mines, holidays, and breakdowns.

Some of the seasonal variation in sulphur shipments stems from the highly seasonal nature of fertilizer consumption. As farmers make most of their purchases in the late spring and early summer, fertilizer manufacturers, who currently account for about half of Frasch sulphur sales, usually operate at full capacity during March, April, and May. Sulphur shipments are therefore higher during these months.

Table 14.   Monthly   Indexes   of   Seasonal   Variation   in   Frasch   Sulphur   Inventories, Production, and Shipments, 1951–68

Month	Inventories	Production	Shipments
January	1.010	1.001	0.922
February	1.011	0.915	0.903
March	1.005	1.012	1.052
April	0.996	1.001	1.059
May	0.986	1.024	1.098
June	0.983	0.991	0.998
July	0.998	1.019	0.950
August	1.003	1.027	0.992
September	1.004	0.965	0.948
October	1.002	1.016	1.039
November	1.002	0.992	0.988
December	1.000	1.037	1.051

SOURCE: Computed from data in U.S. Department of Commerce, Office of Business Statistics, *Survey of Current Business* (1952–69).

Seasonally adjusted monthly Frasch sulphur production, end-of-month inventories, and shipments appear in figure 8. The striking feature of this chart is the stability of inventories and their size relative to output and shipments. Over the period from 1951 to 1968, the ratio of end-of-month inventories to monthly shipments has averaged 8.6 ranging from 15.7 in July 1958 to 2.5 in December 1967. While inventories never fell below 1.9 million tons, shipments never exceeded 866,000 tons.

Prior to the introduction of liquid storage of sulphur in 1959, the size of Frasch inventories could be explained by the ease with which sulphur could be stored in dry bulk form. Between 65 and 70 per cent of the inventory maintained by Frasch producers was being held by Texas Gulf Sulphur Company, even though this firm was accounting for only 40 to 45 per cent of total domestic Frasch production. Texas Gulf's management contended that large inventories had been maintained for two purposes: to achieve mining efficiency by maintaining sulphur production at a reasonably level rate; and to serve as a form of risk reduction for both producers and consumers.[1]

1. See Texas Gulf Sulphur Company, *Annual Report*, for 1956 and 1957. In testimony before the Federal Trade Commission, industry representatives stressed "that in the mining of sulphur, the process of 'steaming' wells is a continuous operation that normally is not interrupted because interruption increases production cost and may mean lessened potential recovery from a given operation. For this reason, wells are normally operated continuously until exhausted and their production accumulates above ground as unsold inventory." Federal Trade Commission, *Report on the Sulphur Industry and International Cartels* (Washington, D.C.: U.S. Government Printing Office. 1947), p. 93.

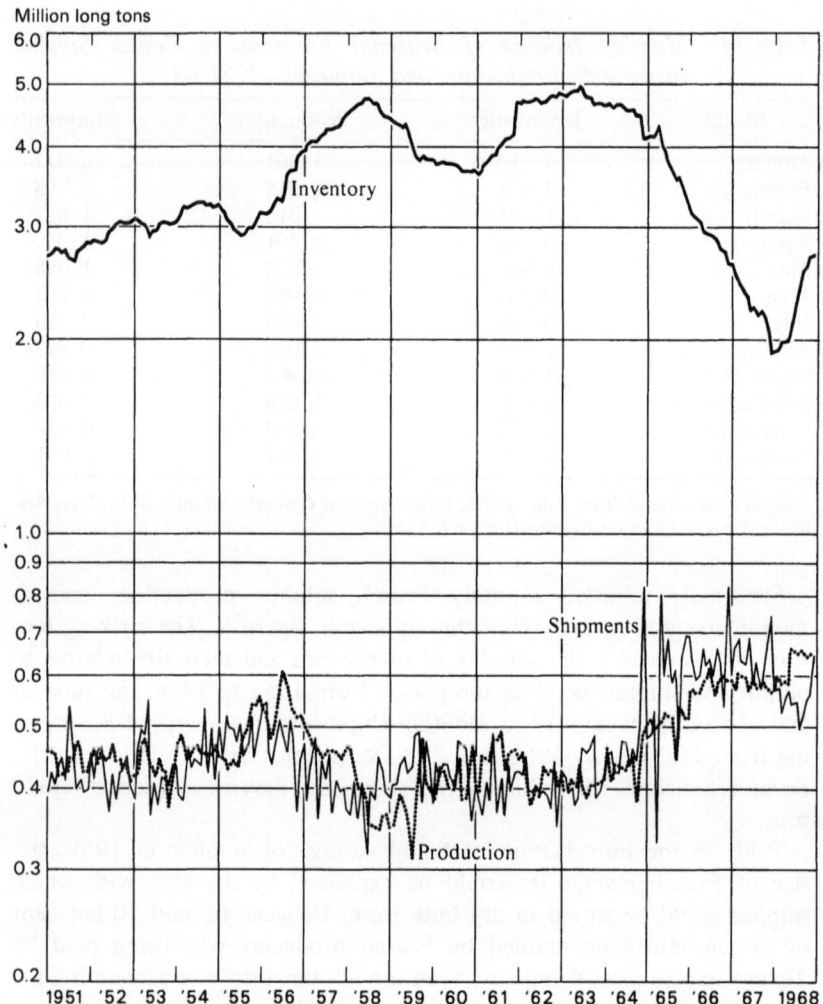

Million long tons

Figure 8.   Monthly U.S. Frasch sulphur production, shipments, and inventories, 1951–68, seasonally adjusted. (Computed from data in U.S. Department of Commerce, Office of Business Statistics, Survey of Current Business, 1952 to 1969.)

Since 1959, two forces have worked to reduce the size of inventories held by producers. The first was the switch to liquid distribution of sulphur which made storage of sulphur more costly and led to cutbacks in the inventories of consumers as well as producers. The second and more important force was the expansion of demand in recent years, which outstripped production and resulted in a steady decline in the industry's stocks from 1963 through 1967. At the end of 1967, producers' inventories amounted to 1.9 million tons, an all-time low for the industry.

The relationship between production and shipments shown in figure 8 is fairly close, particularly for the period prior to 1956. As might be expected, production appears to have fluctuated less than shipments, and the monthly changes in production have been consistently less than the monthly changes in shipments. As Frasch sulphur is generally sold on long-term contracts, producers have a fairly good idea of how much sulphur they will supply over the year, but they cannot be certain of the monthly distribution of shipments. It is not surprising that production tends to suggest the trend about which the shipments series fluctuates.

Trend in annual U.S. Frasch sulphur production

Annual domestic Frasch sulphur production from 1905 to 1967 is shown in figure 9. In order to depict the long-run movement of Frasch sulphur production, the method of least squares was used to derive a

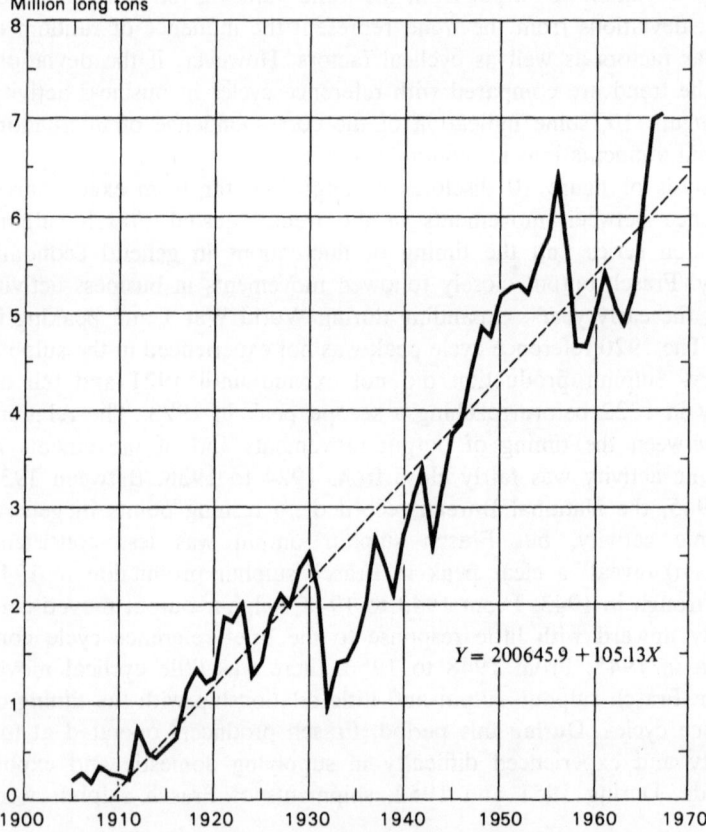

Million long tons

$$Y = 200645.9 + 105.13X$$

Figure 9. Annual U.S. Frasch sulphur production, 1905–67. (Based on data in table A-1.)

linear trend function for this period, depicted as the dotted line in the chart. The coefficient of determination for this trend was 0.8953.

Figure 9 indicates a nearly continuous upward trend in Frasch sulphur production from 1905 to 1967. Output was generally above the trend values until 1930 and below them from 1930 to 1945. In the post-World War II period, output again exceeded the trend values until 1957. The downturn in production from 1958 to 1964 is of interest because, unlike departures from the long-run trend in earlier years, it cannot be attributed to cyclical fluctuations in industrial activity. The slump during these years was the direct result of increased competition from other sources of sulphur.

Cyclical behavior of Frasch sulphur production

One method of evaluating the sensitivity of Frasch sulphur production to cyclical fluctuations in economic activity is to compute the percentage deviation of output from the trend values given in figure 9. Of course, deviations from the trend represent the influence of random or irregular factors as well as cyclical factors. However, if the deviations from the trend are compared with reference cycles in business activity, as in figure 10, some indication of the correspondence of fluctuations in output to fluctuations in economic activity can be gained.[2]

Analysis of figure 10 discloses a rough, but far from exact, correspondence between movements in the trend-adjusted Frasch sulphur production series and the timing of fluctuations in general economic activity. Frasch output closely followed movements in business activity during the early years, expanding during World War I and peaking in 1918. The 1920 reference cycle peak was not experienced in the sulphur industry. Sulphur production did not expand until 1921 and fell off slightly in 1922 before reaching a second peak in 1923. The relationship between the timing of output movements and of movements in economic activity was fairly close from 1924 to 1938. Between 1938 and 1945, the National Bureau recorded no turning points in general economic activity, but Frasch sulphur output was less consistent. Figure 10 reveals a clear peak in Frasch sulphur production in 1942 and a trough in 1943. From 1943 to 1948, sulphur output moved consistently upward with little response to the brief reference cycle contraction in 1945. From 1948 to 1956, there was little cyclical movement in Frasch sulphur output and little relationship with the timing of reference cycles. During this period, Frasch producers operated at full capacity and experienced difficulty in supplying domestic and export demands. During 1951 and 1952, shipments of Frasch sulphur were

2. Reference cycles taken from *Business Cycle Developments* (September 1966), p. 65.

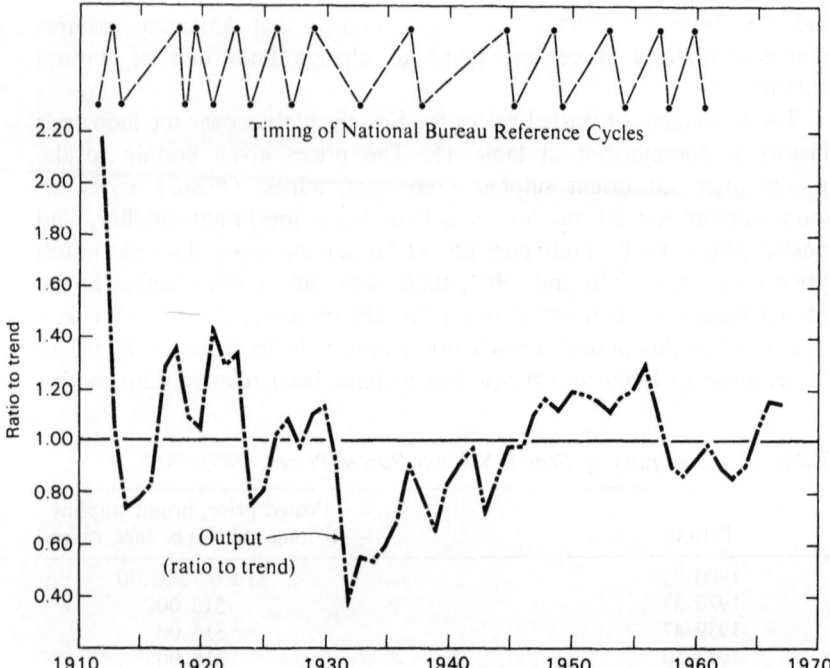

Figure 10.   Annual U.S. Frasch sulphur output—ratios to trend. (Reference cycles from Business Cycle Developments, *September 1966, p. 65.)*

under the control of the Defense Production Agency. The slump in Frasch sulphur production from 1958 to 1963 and the steady increase in Frasch sulphur production since 1964 both appear to have been more closely related to changes in market structure, particularly the entry of new sources of sulphur, than to cyclical movements in economic activity.

Price Behavior in the Frasch Sulphur Industry

Short-run movements in posted Frasch sulphur prices

Frasch sulphur is a perfect textbook example of an industry in which prices are administered in the sense given it first by Gardiner C. Means as "prices . . . made by administrative decisions influenced to a greater or lesser extent by market conditions."[3] In his study of the fertilizer

3. See National Resources Committee, *The Structure of the American Economy*, Pt. I (Washington, D.C.: U.S. Government Printing Office, 1939), p. 109. Means contended that inflexible prices were the result of industrial concentration, and Frasch sulphur was one of the examples given. For a critical review of the arguments for and against the theory of administered prices see John M. Blair, "Means, Thorp, and Neal on Price Inflexibility," *Review of Economics and Statistics* (November 1956) ʾp. 427–35.

industry, Jesse Markham noted: "It is unlikely that American business annals contain a price less given to change than that of natural sulphur."[4]

The movement of posted prices for Frasch sulphur over the industry's history is summarized in table 15. The prices given pertain to the posted price for bright sulphur f.o.b. cars, mines. (Posted prices for dark sulphur are $1 per ton less than those for bright sulphur, and posted prices, f.o.b., Gulf port are $1.50 per ton more than the mines price.) Between 1926 and 1964, there were only seven changes in the posted price—an average of one price change every 5.6 years! Taken over most of this period, Frasch prices appear to have been established in response to long-run criteria and to have been relatively insensitive

Table 15.   Summary of Frasch Sulphur Posted Prices, 1900–1968

Period	Posted price, bright sulphur ($/long ton, f.o.b. cars, mines)
1900–25[a]	$14.00–$22.00
1926–38	$18.00
1939–47	$16.00
1948–50	$18.00
1951–53[b]	$21.00–$24.00
1954–57[c]	$26.50
1958–64[d]	$23.50
1965–66[e]	$25.50
1967–68[f]	$28.00–$42.00

SOURCES: *Oil, Paint, and Drug Reporter;* U.S. Bureau of Mines, *Minerals Yearbook* for the applicable years.

[a] Initial price established by Union Sulphur Company was $17.00. Beginning in 1904, price increased to $18.00 where it remained until 1916. From 1917 to 1919, sulphur sold at $22.00 per ton to munitions and fertilizer manufacturers, although spot prices went much higher. From 1920 to 1926, posted price varied from $14.00 to as high as $22.00 per ton.

[b] In the fourth quarter of 1950, one producer increased his price to $22.00 and the other producer to $21.00 per ton. In 1951, the Office of Price Stabilization froze Frasch sulphur prices at $21.00 to $24.00 per ton.

[c] Price controls were removed in March 1953. Price immediately increased to $25.50 per ton. In 1954, price raised to $26.50 per ton.

[d] In September 1957, price reduced to $23.50 per ton. Between 1958 and 1964, posted price remained unchanged, but producer's realizations on sulphur sales fell to as low as $18.00 per ton due to freight absorption, discounts, etc.

[e] In July 1964, the posted price was increased to $25.50 per ton.

[f] In December 1966, the posted price increased to $28.00 per ton. Posted price increased to $32.00 in April 1967 and to $37.50 in September 1967. In 1968, trade journals were reporting a posted price of $39.00 per ton through March and $42.00 per ton for the remainder of the year.

4. Jesse W. Markham, *The Fertilizer Industry* (Nashville: The Vanderbilt University Press, 1958), p. 79.

to short-run changes in supply and demand. Short-run price stability was maintained by permitting inventories to fluctuate and by adjusting production schedules, when possible, to conditions of demand at the prevailing price.

In the period from 1965 through 1968, the price of Frasch sulphur was increased five times in the face of a severe worldwide shortage of the element. The increased responsiveness of posted prices to short-run conditions in the market represents a relatively recent phenomenon in the industry, but one that is likely to remain. It reflects the increased competitiveness of world sulphur markets since the introduction of new sources of supply in the past decade. The new pattern of pricing means that in periods when sulphur is plentiful, the posted price is likely to become less meaningful as sellers substitute freight absorption, discounts, and other hidden incentives for changes in the posted price. This type of behavior existed in the period from 1958 to 1964 when sulphur markets were glutted. In periods when the market is in short supply, the posted price is likely to increase quite rapidly as it did in the period 1965 through 1968, when price was used as a means of rationing existing supplies among consumers.

Trends and cycles in posted Frasch sulphur prices

Examination of the posted prices for Frasch sulphur given in table 15 reveals no clearly definable trend for the entire period from 1900 to 1968. Several short-run trend elements are identifiable, however, and each is related to a particular phase of the industry's development. A chronological description of industry price and output behavior is given in the next chapter.

Cyclical movements in Frasch sulphur prices are difficult to identify. The price decline in the period following World War I, the abrupt rise in 1926, and the decline in price in 1938 were generally consistent with movements in industrial activity in those years. However, the stable posted price from 1926 to 1937 gives little indication of the severe depression of the early thirties. In the post-World War II period, there appears to have been little relationship between cyclical movements in business activity and Frasch sulphur prices. Frasch prices remained constant during the recessions of 1949 and 1954. The price decline in 1957 preceded the general decline in business activity in 1958, and the posted price of Frasch sulphur remained constant during the 1961 recession. While Frasch prices rose dramatically from 1964 to 1968, the increases were in response to a fundamental imbalance between production and demand, rather than to the cyclical expansion in the economy. Over the entire period, it would appear that cyclical fluctuations in business activity have not affected Frasch sulphur prices. They perhaps have served to reinforce the pressure for revision of the posted price, but with the possible exception of the decline in 1938, they do not appear to have been responsible for the few changes in the posted price.

Price discrimination on the export market

Export prices in the Frasch sulphur industry have always been quoted separately from the domestic price, f.o.b. mines. In the early years of the industry, the export price was quoted on a c.i.f. basis at New York City; in recent years, it has been quoted f.o.b. vessels, Gulf port. Examination of the average values per ton of Frasch exports and Frasch domestic shipments reveals that both have differed significantly from the average annual posted price for Frasch sulphur. Over most of its life, the industry price for export has been higher than the domestic price. Does this constitute discriminatory pricing? Price discrimination may be defined as the "act of selling the same article produced under a single control, at different prices to different buyers."[5] This definition, however, would encompass both the sporadic short-lived discrimination found in many competitive markets and the systematic and persistent discrimination characteristic of monopoly.[6] Only the latter type of discrimination is of interest at this point because it represents the exploitation of market power.

Stigler has established three conditions which are required for price discrimination to exist. First, it must be possible to separate the market into two or more components. Second, demands in the various separable parts of the market must be considerably different. Third, the cost of separating the markets must not be too large.[7] To these conditions, a fourth requirement should be added: uniform pricing policies in the high-price market must be followed by all firms in the industry. The latter condition is essential if prices in the high-price market are not to be pulled down to those in the lower-price market by the actions of the independent marketers.[8]

5. Joan Robinson, *The Economics of Imperfect Competition* (London: Macmillan and Company, Ltd., 1961), p. 179.

6. See M. A. Adelman, "Effective Competition and the Antitrust Laws," *Harvard Law Review* (1948), pp. 1331–32.

7. Stigler, *op. cit.*, pp. 215–16.

8. Joe S. Bain, *The Economics of the Pacific Coast Petroleum Industry* (Berkeley: University of California Press, 1945), vol. II, pp. 195–204, found evidence of persistent and systematic discrimination in the West Coast Petroleum industry between "export" and "domestic" sales of prime fuels because of the existence of the "precise minimum conditions of concentration or of concurrent action which are necessary . . . to make some discrimination possible." On the other hand, Daniel C. Hamilton, *Competition in Oil* (Cambridge: Harvard University Press, 1958), pp. 129–34, found no such evidence of discrimination in the Gulf Coast refinery market and attributed his findings to the lack of concurrent action among sellers.

There are, of course, varying degrees of discrimination. The most common type of price discrimination consists in classifying buyers on the basis of the elasticity of their demand for the product and charging different prices for each class of purchaser. When this type of price discrimination is possible, profits will be maximized by adjusting total output to the level at which marginal cost of the entire output equals the sum of the marginal revenues in the various markets and marginal revenue in each separable market is equal. The analysis assumes that the conditions mentioned above for the existence of price discrimination are met. Over several periods in its history, these conditions have been fulfilled in the Frasch sulphur industry.

With the formation of the Sulphur Export Corporation in 1922, Frasch producers were able to separate their foreign and domestic markets at little or no cost to themselves, and by making all export sales through Sulexco they were able to achieve a uniform pricing policy in the export market.[9] Thus, by establishing a Webb-Pomerene export association, Frasch producers were able to sell at a higher price in export markets.

For discrimination to be profitable, however, the elasticity of demand had to be significantly different in the two markets. There is little reason to believe that at the time Sulexco was formed there was any marked difference between the elasticity of demand for sulphur in the domestic and export markets. Consumers of sulphur, both in the United States and abroad, however, could have been divided into two groups: (1) consumers requiring sulphur in its elemental form (rubber, sulphite wood pulp, insecticide, and pesticide manufacturers, for example), who represented about 20 per cent of the total market for sulphur, and producers of sulphuric acid for medicinal and other purposes requiring pure sulphuric acid which can only be made from elemental sulphur; and (2) manufacturers of sulphuric acid (other than those in the first group) who require sulphur in either elemental or nonelemental form, and who accounted for approximately 75 per cent of total sulphur consumption.

As might be expected, the demand for Frasch sulphur by the first group of consumers was much more inelastic than the demand by the second group of consumers because they could choose only between Frasch sulphur and native Sicilian sulphur. At that time, there was no economical process for recovery of elemental sulphur from gases containing hydrogen sulphide, and the small deposits of native sulphur

9. Canada, Cuba, Newfoundland, and the insular possessions of the United States were not covered by the Sulexco agreement nor by the later agreement with the Sicilian Consorzio. Therefore, there was no formal control over these markets which represented about 15 per cent of the total export market for Frasch sulphur.

that are scattered about the earth could not compete with Frasch sulphur or Sicilian brimstone at anything like the then prevailing price. Acid producers in the second group could have used either of the two forms of elemental sulphur or pyrites or smelter gases as the raw material for their process. Thus, they were much less dependent upon Frasch sulphur than the first group of consumers.

Frasch producers were able to discriminate in the European market by selling their product mainly to consumers in the first group. They further decreased the elasticity of demand in this market by formally agreeing with the Sicilians to divide the world elemental sulphur market (excepting the United States, Italy, Cuba, Canada, Newfoundland, and the insular possessions of the United States) on the basis of 75 per cent to Sulexco and 25 per cent to the Sicilian Consorzio. They cemented their complete control of the world market for elemental sulphur by fixing prices and establishing quotas.

In the United States, however, there was no way for Frasch sulphur producers to separate the domestic market into its component parts. Thus, they had to charge the same price to all domestic consumers, regardless of whether they were restricted to the use of elemental sulphur or not. Since the total demand for Frasch sulphur was more elastic than that for the elemental segment being supplied abroad, the price charged in the United States was lower than that charged for exports.

The above analysis raises two interesting questions, the answers to which shed considerable light on the behavior of Frasch producers. The first question relates to why domestic Frasch producers were willing to concentrate their selling efforts abroad in that portion of the market requiring elemental sulphur and to neglect the much larger acid market. The second question relates to the decision of Frasch producers to compete for the entire market for sulphur in the United States.

Why did Sulexco choose not to actively compete for the 75 per cent of the foreign sulphur market and restrict its main efforts to supplying the portion of the market requiring elemental sulphur? One obvious reason was the existence in Europe of a major combine of pyrites producers who controlled the European acid market. This combine, the European Pyrites Corporation, was owned jointly by the Rio Tinto marketing combine and the Metallgesellschaft, and controlled the marketing of Spanish, Norwegian, Cypriot, and Portuguese pyrites.[10] Much of the European acid market was directly controlled by pyrites producers who owned the principal acid plants. The competitive position of European pyrites was enhanced by the fact that much of the pyrites output was a by-product of copper, lead, and zinc operations. Supported

10. Theodore J. Kreps, *The Economics of the Sulfuric Acid Industry* (Stanford: Stanford University Press, 1938), pp. 100–01.

by such joint products, pyrites were in a virtually impregnable competitive position.

A second reason was the commanding position Sulexco held relative to the only other source of elemental sulphur, Sicily. In 1906, a Sicilian agent reported that the cost of Frasch sulphur landed at a European port was about 20 per cent below the cost of extracting and purifying Sicilian sulphur at the mine. The strong competitive position of Frasch sulphur at the time of the agreement between Sulexco and the Sicilian Consorzio is evidenced by reports which appeared in a leading trade journal in 1921, pointing out that Frasch sulphur could be landed in Europe at 520 lire/ton, while Sicilian sulphur sold for 650 lire/ton at Sicilian ports.[11] The situation in the Sicilian industry was so bad in 1921, that Italy passed legislation forbidding the importation of sulphur.[12] Thus, while Herman Frasch had been content to settle for only one-third of the world elemental sulphur market in his 1907 agreement with the Sicilians, Sulexco was able to demand and receive 75 per cent of the market. In fact, Sulexco was in a position to have eliminated altogether the Sicilian native sulphur industry through price competition, and might have done so had it not felt that the Italian Government would have come to the industry's defense with massive subsidies.

A third reason behind Sulexco's decision not to resort to price competition to gain a major share of the European acid market was the ability of Sulexco to supply some foreign acid markets at the same price that it charged to consumers restricted to using elemental sulphur. The cost of transporting a ton of sulphur in pyrites was over twice that of transporting a ton of sulphur in its elemental form. Therefore, in isolated markets such as Oceania, Asia, and South America, Frasch producers were able to compete with pyrites even at the higher export price. This was also true of some major inland markets in Europe.

Why did Frasch producers decide to compete for the entire sulphur market in the United States? Prior to World War I, Union Sulphur Company had not actively competed for the domestic acid market which was then being supplied by European pyrites. A 1919 study of the United States Tariff Commission concluded that prior to the war, domestic Frasch producers "were able to realize larger profits by maintaining the price of sulphur at $22 per ton f.o.b. New York and supplying the paper and chemical trade rather than by reducing the price of sul-

---

11. Henry L. Geissel, "Italian Sulphur Production Declining," *The Engineering and Mining Journal* (23 July 1921), p. 138, reported that between 1914 and 1920, the cost of producing Sicilian sulphur had risen from 80–85 lire/ton to 420–430 lire/ton. Also, see "Sulphur Crisis in Sicily," *The Engineering and Mining Journal* (17 December 1921), p. 977.

12. Kreps, *op. cit.*, pp. 103–04.

phur to a point where they could obtain the sulphuric acid business."[13]

The disruption of the pyrites markets during the war enabled domestic Frasch producers to gain a dominant position in supplying the domestic acid market. By 1919, nearly 50 per cent of domestic acid was being produced from Frasch sulphur.[14] However, the war had also aided the development of two additional Frasch producers. By 1920, the capacity of the industry was twice the level of prewar demand, stocks were sufficient to last for five years, and the industry was faced with the problem of expanding its sulphur sales. Expansion via Sulexco sales was limited by the need to maintain the high export price, and domestic Frasch producers were able to dispose of only 30 per cent of their production abroad. If they had chosen to restrict themselves to the elemental market at home as well, they would have been able to market only about half of their newly increased productive capacity. It is not surprising then that they adjusted the domestic price to enable them to secure about 60 per cent of the domestic acid market. The reasoning behind the price adjustment made between 1919 and 1926 is discussed in detail in the following chapter.

In the period from 1922 to 1940, price discrimination in the Frasch sulphur industry thus hinged on complete control of the world elemental sulphur market and close cooperation between the domestic producers of Frasch sulphur regarding export pricing. The extent of actual discrimination varied over the life of the industry depending upon the degree to which these two requirements were fulfilled.[15]

---

13. U.S. Tariff Commission, *Information Concerning the Pyrites and Sulphur Industry* (Washington, D.C.: U.S. Government Printing Office, 1919), p. 8.

14. Kreps, *op. cit.*, p. 105.

15. Truett, *op. cit.*, pp. 141–47, offers an alternative explanation of the persistent discrimination against the export market. Citing a case first mentioned by Jacob Viner in *Dumping: A Problem in International Trade* (Chicago: University of Chicago Press, 1923), p. 7, Truett contends that the lower domestic price of Frasch sulphur resulted from a "fear of utilizing monopoly power." This explanation would perhaps account for continued discrimination in the export market following the dissolution of Sulexco in 1952.

Chapter 6

Analysis of Behavior in the Frasch Sulphur Industry

In this chapter an attempt is made to relate the observed market behavior of Frasch sulphur producers to changes in the industry's market structure. For this purpose, the development of the industry is divided into three periods. The first period, 1900 to 1926, covers the birth of the Frasch sulphur industry and its growth to dominance of the world sulphur market. In the second period, 1927 to 1946, the industry is characterized by a stable market structure centered about an international cartel. The final period, 1947 to 1968, follows the dissolution of the international sulphur cartel and features the introduction of new sources of supply of both Frasch and other forms of elemental sulphur and their impact on world sulphur markets.

1900 to 1926: Formative Years

Entry of Union Sulphur Company

When the Frasch sulphur industry emerged in the first decade of this century, the world elemental sulphur market was being completely supplied by the native sulphur mines of Sicily, two-thirds of which were under contract to a single firm, the Anglo-Sicilian Company. The United States was one of the best customers for Sicilian sulphur, taking about one-third of the industry's annual output. Competition from Frasch sulphur reduced U.S. imports of Sicilian sulphur from 189,000 tons in 1906 to 20,000 tons in 1907. Falling prices and mounting stocks led to changes in the Sicilian sulphur industry, and in 1906, the Anglo-Sicilian Company was replaced by a cartel, the "Consorzio Obbligatorio per l'Industria Solfifera Siciliana," which was established under Italian law and given total control of all Sicilian sulphur production.

In 1907, after a brief round of price cutting, Union Sulphur Company and the Consorzio reached an agreement regarding the division of the world elemental sulphur market. At that time Union was the only U.S.

Frasch producer; Freeport began production in 1912 and Texas Gulf in 1919. The agreement specified that the world market, excluding the United States and Italy, was to be divided between the two companies on the basis of one-third to Union, and two-thirds to the Consorzio. The agreement established a minimum price for Louisiana sulphur of $22 per ton c.i.f. New York. It also specified that the price differential between Union and the Consorzio was never to exceed 2.5 per cent.[1]

Although the agreement with the Consorzio remained in effect for only four years, the price stability achieved through this accord lasted until after the outbreak of World War I. From January 1909 until February 1916 the posted price of Frasch sulphur remained constant at $22 per ton f.o.b. New York and $18 per ton f.o.b. the mine.

The impact of World War I

Between 1907 and 1915, the Union Sulphur Company had a virtual monopoly of the domestic market for elemental sulphur. During this period, Union concentrated on selling to the paper-pulp and other chemical industries which represented growing markets for elemental sulphur. No serious attempt was made to compete for the domestic sulphuric acid market which had been committed to pyrites since before the turn of the century.

During the First World War the domestic acid market opened up for Frasch sulphur when American sulphuric acid manufacturers were cut off from their supply of Spanish pyrites. While only 2.6 per cent of the acid produced in the United States in 1914 was made from Frasch sulphur, by 1918, approximately 48 per cent of the domestically produced acid was made from Frasch sulphur.[2] In 1916, the contract price of Frasch sulphur increased from $18 to $22 per ton f.o.b. mines. The price of sulphur to munitions and fertilizer manufacturers was voluntarily maintained at that level until 1 January 1919, although the spot price of sulphur in the New York market rose to $45 in the second quarter of 1917.[3]

The opening of the sulphuric acid market to Frasch sulphur and the increased price of sulphur during the war hastened the development of production by Freeport Sulphur Company and Texas Gulf Sulphur

1. *Investigations of Concentration of Economic Power,* Hearings before the Temporary National Economic Committee of the 76th Congress (1939), Part U, pp. 2219–26. (Hereinafter cited as TNEC Hearings.)

2. A. E. Wells and D. E. Fogg, *The Manufacture of Sulphuric Acid in the United States,* U.S. Department of the Interior, Bureau of Mines, Bulletin 184 (1920), p. 27.

3. United States Tariff Commission, "Industrial Readjustments of Certain Mineral Industries Affected by the War," *Tariff Information Series,* No. 21 (1920), pp. 238–39.

Company. With the end of the war, the use of Frasch sulphur to supplement pyrites in the manufacture of sulphuric acid ceased almost immediately. "During the spring of 1919, brimstone stocks at acid plants were gradually used up, and acid manufacturers were inclined not to replenish their stocks until it was apparent that brimstone could be purchased on a parity basis with pyrites ore, the importation of which was gradually resumed."[4] To make matters worse, 1920 ushered in a severe business recession, both in the United States and abroad, and the demand for sulphur declined.

The three domestic Frasch producers, faced with surface inventories nearly five times annual prewar consumption and current production nearly twice annual prewar demand, were primarily concerned with maintaining or increasing their wartime share of the domestic sulphur market. To do so, they had to adjust their prices to meet competition from both domestic and imported pyrites. The price adjustment between brimstone and pyrites which occurred in the seven years following 1919 is of much significance for it established the pattern of the industry price behavior for the next twenty years.[5]

Postwar price adjustments

From 1900 to 1916, with the price of sulphur at $22 per ton (22 cents per unit) f.o.b. New York and the price of pyrites at 13 cents per unit c.i.f. New York, 98 per cent of the sulphuric acid produced in the United States was made from pyrites.[6] A 1920 study by the U.S. Bureau of Mines concluded that if the consumer of pyrites received nothing for the by-product sinter, Frasch sulphur was worth to him only 3 to 4 cents per unit more for acid manufacturing purposes than sulphur in pure pyrites containing 40 to 43 per cent sulphur.[7]

Under these conditions, and with three firms producing a homogeneous product, it would be expected that the price of Frasch sulphur would have to fall if U.S. producers were to maintain their share of the domestic sulphuric acid market in the postwar period. The price of sulphur did fall from $45 per ton in 1917 to $14 per ton in 1922, the lowest point it had reached since 1906. For the next three and one-half years, the price remained at this level. Though pyrites prices fell during

4. Wells and Fogg, *op. cit.*, p. 24.
5. Much of what follows in the way of analyzing the behavior of Frasch prices from 1917 to 1926 is based on a study by Thurmond L. Morrison, "The Economics of the Sulphur Industry" (Ph.D. dissertation, Economics Department, University of Texas, 1939), pp. 100–20
6. A unit is one per cent of a ton (22.4 pounds in the case of a long ton). Thus a long ton of pyrites containing 45 per cent sulphur would contain 45 units (1,008 pounds) of sulphur.
7. Wells and Fogg, *op. cit.*, p. 24.

the same period, the percentage of sulphuric acid produced from Frasch sulphur increased to 75 per cent, as elemental sulphur almost entirely displaced imported Spanish pyrites in the manufacture of sulphuric acid. Much of the remaining 25 per cent of sulphuric acid production not being produced from elemental sulphur was supplied by domestic pyrites producers as a by-product of copper, lead, or zinc operations. In addition, some of the acid plants in the United States were owned by either domestic or foreign pyrites producers. Thus, decreases in the price of elemental sulphur below $14 per ton could not be expected to increase the share of the sulphuric acid market held by domestic Frasch producers because the remaining portion of the market was tied directly to pyrites consumption and insulated from competitive pressures.

It does not follow that $14 per ton was necessarily the price that maximized the industry's profit. Above $14 per ton, increases in the price of sulphur would cause a loss in sales because domestic acid producers would switch from elemental sulphur to pyrites. At a price elasticity less than unity, the loss in tonnage could have been more than compensated for by an increase in total revenue resulting from the higher price. Indeed, if the elasticity of the industry demand curve above $14 per ton were known, the most profitable output and price combination of the industry could be determined. However, there was no way for the firms in the industry, acting independently, to determine the elasticity of the industry's demand at such prices. Under the conditions described, any firm that increased its price above $14 per ton stood to lose all its sales.

In view of the homogeneous nature of the product, rival firms might be forced to follow a price decrease, but they would be reluctant to follow a price increase. Thus, once decreases in the price of sulphur were halted at the level of $14 per ton by the realization that further declines in the price would not increase sales, uncertainty as to a rival's response probably reinforced this price and led to its being maintained at this level.

In October 1922, the three American companies set up the Sulphur Export Corporation under the provisions of the Webb-Pomerene Act, which Congress had passed in 1918. One of the first acts of the corporation, which was formed to handle all export sales of Frasch sulphur, was to negotiate an agreement with the Sicilian producers through the Consorzio. The agreement, which was subject to renewal every four years, provided for the division and allocation of the world sulphur markets, with the exception of Italy and North America, on the basis of 75 per cent to the Sulphur Export Corporation and 25 per cent to the Consorzio. Prices, terms, and conditions of sale of all sulphur sold under the agreement were to be fixed from time to time by the parties in a manner that would best serve their mutual interest. To facilitate control over the world sulphur markets, each party was to furnish to the other party and to a central bureau a monthly statement showing the total tonnage shipped, total tonnage sold, the total tonnage delivered, destinations, prices realized, freight rates, and other such information as was from time to time deemed necessary for "proper forecast and allocation." A penalty of two tons was stipulated for each ton shipped in violation of the agreement. Finally, the two parties agreed that "the situation of the sulphur manufacturing industry in the countries covered by the agreement should be maintained throughout the life of the agreement." The agreement stated that "each party agrees not to do or encourage anything which would result in altering such present situation and any action of a nature to alter such present situation shall be jointly considered and both parties shall use their best endeavors to prevent any such alteration."[8]

The formation of the Sulphur Export Corporation and its subsequent agreement with the Sicilian Consorzio put an end to independent decisions by the three domestic Frasch producers on prices and output for the export market. It has been contended that the domestic Frasch producers used this newly created opportunity for formal cooperation to study the nature of the domestic demand for sulphur at prices above $14 per ton in order to determine the price that would be most profitable.[9] The subsequent behavior of pricing in the industry supports this judgment.

Within eight months of the date on which Sulexco reached an agreement with the Sicilians, the domestic price for Frasch sulphur began to rise. By 1925, the price of Frasch sulphur had risen to $18.50 per ton f.o.b. the mines, $22.00 per ton delivered in New York, Baltimore, and Philadelphia, and $22.50 per ton in Portland, Maine. At this time, imported Spanish pyrites were selling for 13 cents per unit (one per cent of a ton) in these markets.[10] As previously mentioned, the parity

8. The agreement is contained in its entirety in the TNEC Hearings, pp. 2214–17.
9. This view is expressed by Morrison, *op. cit.,* p. 109, *et passim,* pp. 110–15. Similar conclusions were reached by Markham, *The Fertilizer Industry,* pp. 80–83, and p. 178; and by George W. Stocking and Myron W. Watkins, *Cartels or Competition?* (New York: Twentieth Century Fund, 1948), pp. 260–61. The U.S. Federal Trade Commission, *loc. cit.,* pp. 14–16, concluded: "The domestic and export segments of the American sulphur industry are inseparable in interest. From an economic standpoint, the distribution and pricing activities of Sulphur Export Corporation had a natural relationship to the production, distribution, and pricing activities of its individual producing members."
10. Kreps, *op. cit.,* p. 103. The TNEC Hearings (p. 2203) give the price of imported pyrites at 11.5 cents per unit of sulphur in 1925. However, this includes the lower-valued Canadian imports as well as the Spanish imports.

prices for Frasch sulphur and pyrites, as established by the relative cost of manufacturing sulphuric acid from either material, were estimated in about 1920 to be three to four cents higher per unit of elemental sulphur than per unit of pyrites. Given the price of 13 cents per unit for imported Spanish pyrites and about 22 cents per unit for Frasch sulphur, both delivered to the major East Coast markets, it was to the advantage of acid producers in these markets to substitute imported pyrites for Frasch sulphur. Consequently, 85 per cent of the acid producers on the East Coast used pyrites.[11]

In inland markets, however, the relative costs of transporting Frasch sulphur and pyrites were the determining factor in the selection of a raw material for the manufacture of sulphuric acid. The cost of transporting a ton of sulphur in pyrites was over twice that of transporting a ton of sulphur in elemental form.[12] Thus, on shipments from the Atlantic Seaboard to inland markets any differential in price in favor of pyrites was quickly overcome by the freight differential.

It has been estimated that domestic consumption of sulphur fell from 1.6 million tons when the price of Frasch sulphur was $14 per ton f.o.b. the mines to 1.4 million tons when the price was raised to $18 per ton f.o.b. the mines; however, gross revenue rose from $22.4 million to $25.2 million. Even if it had cost nothing to produce the additional 200,000 tons of sulphur, it would still have been unprofitable for domestic Frasch producers to have lowered their price sufficiently to have secured the Atlantic Coast acid trade.

Increasing the price of Frasch sulphur above $18 per ton f.o.b. the mines, however, would have jeopardized the interior acid markets of Western New York, Indiana, Pennsylvania, Ohio, and Illinois, which represented about 30 per cent of the total domestic acid market. For example, it has been estimated that had the price of Frasch sulphur increased to $20 per ton f.o.b. the mines, at least 300,000 tons of sulphur sales would have been lost, causing gross sales revenue to decline from $25.2 million to $22 million.[13]

It was the possibility that acid makers might substitute pyrites for Frasch sulphur which exerted the greatest influence in determining domestic Frasch sulphur prices. As it was unprofitable for acid producers to substitute pyrites for sulphur as long as the price of sulphur was less than $14 per ton f.o.b. the mines and the price of pyrites was 12 to 13

11. Andrew M. Fairlie, *Sulfuric Acid Manufacture* (New York: Reinhold Publishing Corporation, 1936), p. 83. Actually, a larger proportion probably would have switched to pyrites had it not been for the fact that sulphuric acid for medicinal and other special purposes must be made from elemental sulphur.

12. About 45 per cent of the weight of pyrites, as opposed to over 98 per cent of the weight of elemental sulphur, becomes available for acid manufacture.

13. Morrison, *op. cit.*, pp. 112–14.

cents per unit, the demand for sulphur up to this point was quite inelastic. As the price of Frasch sulphur was increased above $14 per ton f.o.b. the mines, demand became increasingly more elastic. Apparently $18 per ton f.o.b. the mines was the price at which Frasch producers felt that their profits were at a maximum, i.e., the price at which the marginal cost of producing sulphur was equal to marginal revenue.[14]

1927 to 1946: Years of Stability

Early in this period the Union Sulphur Company was forced to retire from the industry owing to depletion of its Louisiana mine and its failure to secure additional production.[15] Texas Gulf and Freeport agreed to continue Sulexco and to share equally in its operation.[16] Two new domestic Frasch producers entered the industry—Duval Texas Sulphur Company in 1928, and Jefferson Lake Oil Company in 1932. In 1930, Orkla Grube Aktiebolag, a Norwegian pyrites producer, built a pilot plant in Norway to produce elemental sulphur from pyrites, and licensed the leading Spanish pyrites producer, Rio Tinto, and the leading Portuguese pyrites producer, Mason and Berry, to use its process in their respective countries. About the same time, the Montecatini interests in Italy developed sulphur production from pyrites outside the control of the Consorzio and began selling competitively in France and other markets covered by the Sulexco-Consorzio agreement. The net effect of these developments was to introduce new and, at the time of their development, unpredictable competition both in the domestic market and abroad.

The agreement between Sulexco and the Sicilian Consorzio became inoperative in 1932 because of the dissolution of the Consorzio by the Italian Government. The Sicilian sulphur industry, with the aid and encouragement of the Italian Government, was subsequently reorganized

14. It has been pointed out that there are many obstacles which prevent joint profit maximization even in an oligopoly producing a completely homogeneous product under conditions of identical and horizontal average total cost functions for each firm in the industry. If these latter conditions are relaxed, the problem becomes even more difficult, with joint profit maximization requiring interfirm transfers for the simultaneous maximization of the profits of each firm. See William Fellner, *Competition Among the Few* (New York: Alfred A. Knopf, 1949), pp. 120–36. Given the homogeneous nature of sulphur, and the existence of only two firms in the industry, $18.00 per ton f.o.b. the mines may be interpreted as the price which tended to maximize profits for both producers.

15. The following account of behavior in the Frasch sulphur industry during this period is based on information in the TNEC Hearings, and the Federal Trade Commission, *Report on the Sulphur Industry and International Cartels*, 1947 (hereinafter cited as the Federal Trade Commission Report).

16. A copy of this agreement between Texas Gulf and Freeport is contained in the TNEC Hearings, pp. 2235–36.

as Ufficio per la Vendita dello Zolfo Italiano. On 1 August 1934, an agreement containing provisions similar to the original Sulexco-Consorzio agreement was reached between Sulexco and Ufficio.[17]

Even with these major changes in the composition of the Frasch sulphur industry and its relationship with the Sicilians, which coincided with the severe worldwide depression of the thirties, the period from 1927 to 1946 was marked by an extremely rigid posted price for Frasch sulphur which changed only once during these two decades, and by consistently higher realizations on exports than on domestic sales. The practice of price discrimination, as emphasized earlier, depended upon the ability of Sulexco, through its agreement with the Consorzio, to control the supply of elemental sulphur. By exploiting this control, however, Sulexco encouraged both research aimed at lessening the dependence of foreign consumers upon the cartel and development of new sources of supply. The Federal Trade Commission aptly noted:

> The problem faced by the cartel participants (Sulexco and the Consorzio) was how to obtain the cartel objectives of price protection for the benefit of both Italian and American natural sulphur producers, without disturbing the existing status of the industry by fixing foreign market prices so high as to further stimulate increased production by independent producers of both natural and by-product sulphur.[18]

During the late 1920s, as Europe began to experience the severe depression which was to sweep the world during the next decade, Frasch sulphur export prices advanced, realizations on Frasch sulphur exports rising from $18.92 in 1926 to $21.68 in 1931. By the early thirties, developments on several fronts began to threaten the cartel's control of world markets for sulphur.

Entry of two independent U.S. producers

The entry of Duval in 1928 and Jefferson Lake in 1932 expanded the capacity of the domestic Frasch industry in the thirties when both domestic and export markets were shrinking. Since neither firm became a member of Sulexco, the entry of these two new firms posed a potential threat to Sulexco's cartel agreements covering export markets and to the domestic price stability that had existed in the industry since 1926. The pattern of domestic production and exports from 1932 to 1940 clearly reveals that Texas Gulf and Freeport met these threats by (1) cutting back on their production, permitting Duval and Jefferson Lake to gain a significant share of total production, and (2) through Sulexco aiding Jefferson Lake and Duval to find export markets, even though this meant relinquishing part of their own export sales.

17. A copy of this agreement is contained in the TNEC Hearings, pp. 2208–13.
18. Federal Trade Commission Report, p. 68.

Table 16. Production of Frasch Sulphur, 1930–40

Year	Output	Production of Sulexco members			Production of nonmembers	
		Texas Gulf	Free-port	Total	Duval	Jefferson Lake
	thousand long tons	(................ *per cent of total* ....)				
1930	2,559	67.88	30.58	98.46	1.51	—
1931	2,129	60.95	37.42	98.44	1.63	—
1932	890	63.79	32.04	95.83	2.54	1.50
1933	1,406	46.13	29.45	75.58	2.70	21.61
1934	1,421	59.48	31.72	91.20	3.14	5.33
1935	1,633	56.55	37.54	94.09	3.94	1.50
1936	2,016	63.80	29.70	93.50	5.89	0.42
1937	2,742	63.60	25.95	93.55	4.82	3.44
1938	2,393	51.08	28.59	79.67	8.87	11.20
1939	2,019	39.14	37.83	76.97	12.97	9.95
1940	2,732	52.28	31.49	83.77	7.96	8.04

SOURCE: Jesse W. Markham, *The Fertilizer Industry* (Nashville, Tennessee: The Vanderbilt University Press, 1958), p. 77.

NOTE: Percentages may not add to 100 because of rounding and the inclusion of some non-Frasch sulphur in the output figure.

The adjustments in production made by Texas Gulf and Freeport in response to the entry of Duval and Jefferson Lake can be seen in table 16. From 1930 to 1934, the market share of the two major producers fell from 98 per cent to 76 per cent. The temporary increase in their market shares from 1935 to 1937 reflected the reduction in output of the two independent producers due to exhaustion of their initial mines. After 1937, the market share held by the two major producers again fell, as Jefferson Lake and Duval began exploitation of new deposits. Throughout the period from 1932 to 1935, shipments exceeded industry production as Texas Gulf and Freeport drew on large inventories accumulated prior to and during the early depression years. Thus, Texas Gulf and Freeport were able to maintain price stability in the domestic market during the 1930s by reducing their production, thereby permitting the two new entrants to dispose of their production at the existing price of $18.00 per ton.

The bulk of the production from the initial deposits exploited by Duval and Jefferson Lake came onto the market in 1933 and 1934, at a time when the agreement between Sulexco and the Consorzio was inoperative. Nevertheless, the price structure in European markets remained essentially the same as it had been under the cartel agreement. During these two years it became imperative that Sulexco obtain some control over the exports of the two independents or else face a disruption of the foreign price structure. The matter of exports by independent U.S. Frasch producers also became a factor in negotiations with the newly created Ufficio.

In 1933, Sulexco entered into direct agreements with Duval and Jefferson Lake to allocate export markets to these firms in return for their pledge to keep export prices intact, to limit their exports, and to confer with Sulexco as to destinations and prices before accepting any orders.[19] Similar agreements were made in 1934. As a result of these agreements, Sulexco marketed 86,519 tons for Jefferson Lake and 10,000 tons for Duval in European markets in 1933. In 1934, the two independents exported 117,216 tons, 110,700 tons of which were sold under agreement with Sulexco. Thus, nearly 94 per cent of the 125,128 tons produced by the two independents in 1934 was marketed abroad, leaving only 7,912 tons to be sold competitively either in the domestic market or elsewhere. In its 1934 agreement with Ufficio, Sulexco agreed to deduct from its quota any export shipments by independent American producers to the joint territory.

The record of cooperation between Sulexco and the two independent American Frasch producers in marketing sulphur abroad provides a good insight into the behavior of Frasch output and prices in the thirties. When exports of the American independent producers threatened to disrupt the foreign price structure and the negotiations with the Italians for a new cartel contract, Sulexco reacted by aiding Duval and Jefferson Lake in finding foreign markets, even though this meant relinquishing possible sales for its own members. Sulexco's actions during this period in which there was no international agreement in effect with the Sicilians enabled it to protect its foreign price structure, while at the same time removing from possible competition in the domestic market a corresponding quantity of independent tonnage. Thus, an attitude of cooperation based on mutual advantage was established in 1933 and 1934 between Sulexco members and nonmembers with respect to export trade.

Such cooperation was clearly to the advantage of Duval and Jefferson Lake. With export prices higher than domestic prices, and with Sulexco assisting Duval and Jefferson Lake to place export sales, neither of the two independents could have hoped to have reaped additional advantage from competing sharply in price with Sulexco members in the domestic market. To have done so would not only have reduced their realizations on sales in the domestic market, but would also have endangered their allocations in foreign markets.

---

19. Copies of these agreements are contained in the TNEC Hearings, pp. 2236–37.

After the renewal of the international agreement between Sulexco and the Italians, it appears that somewhat similar policies, again based on mutual advantage, were followed both at home and abroad. The large companies appear to have avoided using their superior economic strength and position in the market to compete with nonmembers in price either at home or abroad, because they appear to have realized that price competition with Duval and Jefferson Lake might have destroyed their cartel arrangements in Europe. With realizations on exports running $3 to $4 a ton above realizations on domestic sales, both Freeport and Texas Gulf had a vital interest in maintaining a stable price structure. With two larger firms in the industry taking this attitude, it would not have been wise for either Duval or Jefferson Lake to have initiated a price conflict.

Even though no arrangements existed between Sulexco and the two independent producers in regards to export markets after 1934, the two smaller firms were closely related to Texas Gulf Sulphur, the largest American producer. During the late thirties, after its initial mine had been depleted, Jefferson Lake obtained options from Texas Gulf on the Clemens Dome which it operated from 1937 to 1960. It paid sulphur royalties to Texas Gulf from 1937 to 1949. Duval operated a portion of Boling Dome by royalty agreement with Texas Gulf and obtained its options on Orchard Dome from Texas Gulf.

Diversion from domestic to export markets

Other developments both at home and abroad challenged Sulexco's position. The differential between domestic and export prices for Frasch sulphur encouraged American purchasers for domestic consumption to divert sulphur to the export market. Such diversions outside the control of the cartel became a disturbing factor to the cartel participants. American sulphur producers had always handled their own export sales and avoided setting up independent American exporters of crude sulphur either in the capacity of independent export merchants or as local agents for foreign buyers. Under the Sulexco agreement, Texas Gulf and Freeport agreed to turn over to the Corporation all orders or inquiries for sulphur to be shipped to the marketing territory covered by the agreement. In general, requests by nonconsumers for purchases for foreign producers were declined by Sulexco.

In the late 1920s, Sulexco found that its control of exports from the United States was being undermined by exports of domestic firms who were purchasing their sulphur from Sulexco members and diverting it to foreign markets. On 27 May 1929, Freeport and Texas Gulf sought to remedy this situation by making a supplemental agreement to their Sulexco contract. The proposed change read:

No exporter shall sell to a domestic purchaser more crude sulphur than is necessary for the domestic sales requirements of such purchaser, if such sales would in any way interfere with or adversely affect the business of the exporters as carried on by the Sulphur Export Association or tend to deprive them of the benefits of the Webb Act.[20]

Upon submission of the proposed supplemental agreement to the Federal Trade Commission, Freeport and Texas Gulf were informed by the Commission that the clause represented an obvious attempt at restraint of trade. On 16 October 1929, the parties filed with the Commission a formal abrogation of the agreement, stating that no action had ever been taken under it.

The Norwegian episode

Still another challenge to Sulexco's dominance of the world elemental sulphur market came from the development of a process in Norway to produce elemental sulphur from pyrites. Prior to 1931, Texas Gulf Sulphur, realizing the potential of processes designed to extract elemental sulphur from pyrites, had conducted experiments with various processes, accumulating twenty-five to thirty patents. In 1930, Texas Gulf obtained an exclusive license for North and South America to use a process developed by a Norwegian firm, Orkla Grube Aktiebolag, for producing sulphur from pyrites.

Orkla itself began production on a commercial scale in 1931, and sold 7,600 tons in 1931 and 53,800 tons in 1932 to consumers in Scandinavia at prices less than the prevailing c.i.f. prices of the cartel. In January 1933, Sulexco made an agreement with Orkla, whereby Orkla would restrict its output to a maximum of 70,000 tons per year to be sold in Scandinavia at a minimum price agreed upon by both Sulexco and Orkla; in return, Orkla would receive a payment of $1 per ton by Sulexco for each ton of Orkla production up to 70,000 tons annually.[21] Thus, Sulexco was able to protect its European markets from competition with Orkla. Sulexco renewed its 1933 agreement with Orkla annually through 1936. Sulexco negotiated a new five-year agreement with Orkla beginning 1 January 1937. The new agreement provided that Orkla could increase its capacity beyond 70,000 tons per year, extended Orkla's market privileges to Europe, Asia, Africa, and adjacent islands, eliminated the tonnage payment by Sulexco, and divided sales in the joint territory on the basis of one-third to Orkla and two-thirds to Sulexco. The agreement provided that "in no case would an attempt be made to secure customers who would be in direct competition with pres-

20. Federal Trade Commission Report, pp. 53–54.
21. Ibid., pp. 68–69.

ent users of brimstone without the mutual consent of both parties."[22] Bolstered by the new agreement, Orkla's production increased to 105,000 tons in 1937 and 108,000 tons in 1938 and 1939.

Sulexco's 1936 agreement with Orkla also stated that it might be desirable to obtain certain patents relating to sulphur production in the joint territory. The cost of acquiring any such patents was to be shared jointly, Orkla paying one-third and Sulexco two-thirds of the cost. The title to patents so acquired was to be held in trust by Orkla's subsidiary, A.B. Industrimetoder, Stockholm. No license for the use of such jointly owned patents would be issued without the consent of both parties. Later testimony before the Federal Trade Commission revealed that the patent purchasing clause was inserted in the 1936 Orkla agreement because one Orkla official knew of certain experiments underway in Norway and feared that the Norwegian Government would finance the development of the patents. Under this agreement, Sulexco contributed $35,000 as its share of the purchase cost of patents purchased by Orkla. Testimony revealed: "The development stopped on them."[23]

Behavior within the Frasch sulphur industry between 1927 and 1940 was aptly summarized by the Federal Trade Commission:

> The international picture, in its entirety, then is that low-cost, privately-owned American producers, organized as a Webb Act Association, combined, on the one hand, with a government sponsored Italian producing and marketing cartel and, on the other, with a privately owned Norwegian patent owner, to fix and maintain prices, divide markets, restrain competition and control sulphur developments in foreign countries. Controls broke down, or threatened to break down, from time to time due to the rise, or threatened rise, of new competition such as that of Orkla interests in the Baltic and other markets, Montecatini interests on the Italian mainland and in France, and that of Spanish, Portuguese, and independent American interests selling, or threatening to sell, into France and other countries. The fact that such failures and threats of failure were promptly met by efforts to rebuild and strengthen the cartel's structure and controls only serve to emphasize the real purposes and effects of the various cartel agreements on industrial developments and commerce in and with countries other than those of the participants.[24]

During the Second World War, all of the agreements between Sulexco and foreign sulphur producers became inoperative. On 13 September 1939, Sulexco gave notice to the Italians of complete suspension of the Ufficio contract effective as of 1 September 1939, because of war. On 15 March 1940 the Italians gave notice to Sulexco of the suspension of all agreements under which they were operating. Finally, pursuant to

22. A copy of this agreement is contained in the TNEC Hearings, pp. 2240–41.
23. Federal Trade Commission Report, p. 71.
24. *Ibid.*, p. 84.

action by its Board of Directors, on 15 February 1945, Sulexco notified Ufficio under date of 21 February 1945 of the cancellation of the agreement of 1 August 1934 as amended and supplemented. On 19 September 1939, Sulexco caused the complete suspension of the Orkla contract because of war conditions in Europe, and the contract was formally cancelled by Sulexco on 22 February 1945.

1947 to 1968: Years of Change

The period from 1947 through 1968 in the Frasch sulphur industry was dominated by successive phases of shortage and surplus in sulphur markets at home and abroad. The first shortage developed in the period following World War II, and culminated in controls being placed on the industry during the Korean crisis. This was followed by the development, beginning in 1953, of alternative sources of supply for elemental sulphur both within the United States and abroad. The introduction of new sources of supply created a surplus in sulphur markets from 1959 to 1963. During this period, the traditional pricing structure of the industry was destroyed, and a complete reorganization of Frasch sulphur marketing took place with the introduction of liquid shipment and regional terminals. The final years, 1964 to 1968, witnessed the reappearance of a shortage in sulphur markets as supply failed to keep pace with surging demand.

The patterns of production and consumption, exports and imports, and domestic and export prices in the sulphur industry over the period from 1947 to 1968 are shown in figures 11, 12, and 13. At the top of each chart the major factors affecting sulphur markets are indicated as well as the state of the U.S. sulphur market. U.S. production and consumption trends show the following pattern over the period: 1947 to mid-1956, shortage (drawing down of inventories); mid-1956 to mid-1959, balance; mid-1959 to mid-1963, oversupply (accumulation of inventories); mid-1963 to mid-1964, balance; mid-1964 to 1967, shortage; and 1968, balance.

Postwar adjustments

The 1947 Federal Trade Commission study of the sulphur industry and the operation of the Sulphur Export Corporation recommended that Sulexco refrain from entering into agreements with foreign producers to establish quotas which included domestic nonmember firms, to establish guarantees of tonnages to foreign producers, or to maintain the status quo in the manufactured sulphur industry; from acquisitions of patents "useful for or capable of being used in connection with the production of sulphur for commercial purposes"; from entering into any

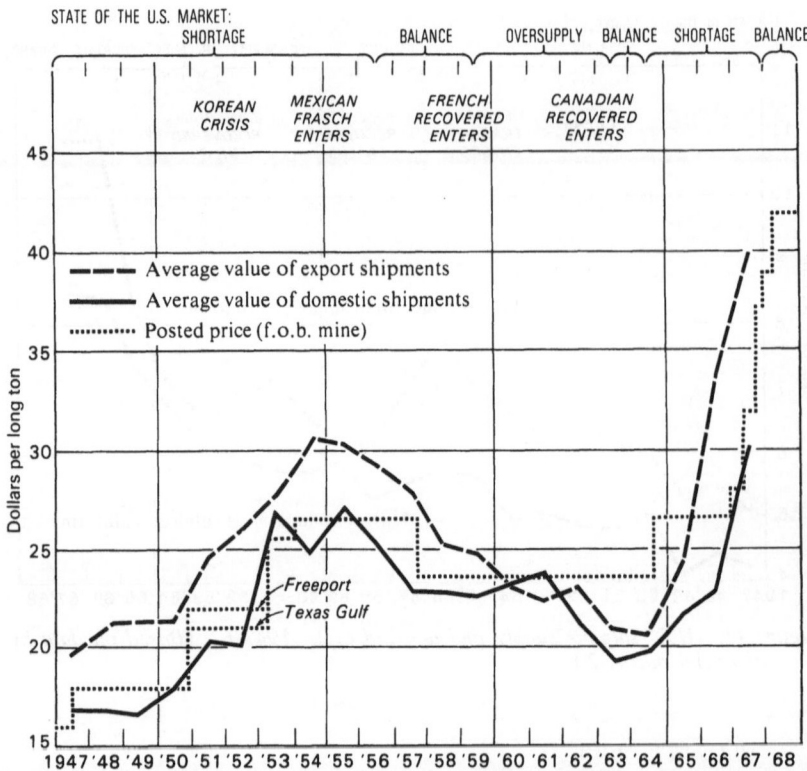

Figure 13. Average value of export and domestic shipments of U.S. Frasch sulphur and posted price, f.o.b. mine, 1947–68. (Based on data in tables 15 and A-3.)

agreement or understanding with nonmember domestic producers; and from selling or assisting in the sale of output from nonmember domestic producers.[25] However, by 1947 these were more or less academic issues because the Frasch sulphur industry had divorced itself from its international agreements during the war, and the Sulexco members had made no further attempt to obtain agreements with either domestic or foreign producers.

As shown in figure 11, the demand for Frasch sulphur increased rapidly in the years following the end of World War II. The largest domestic consumer of sulphur, the sulphuric acid industry, increased its output from 766,800 short tons in 1946 to 1,085,000 short tons in 1950. Exports of crude sulphur, depicted in figure 12, which had averaged 566,361 tons between 1935 and 1939, increased to an average of

25. *Ibid.*, pp. 102–3.

1,366,361 tons between 1947 and 1950. Responding to the exp in demand, Freeport raised its price from $16 to $18 per tor the mine in June 1947, and two months later Texas Gulf met th increase (see figure 13).

Glenn Lehmann, in his study of the industry, has empha effective role played by the Justice Department in the immed war period in preventing additional price increases on Frasch Industry officials had experienced public scrutiny of their op the Temporary National Economic Committee of Congress i by the Federal Trade Commission in 1947, and they were avoid any action that might draw public attention to the nature of the industry. Representatives of the major produ quently during the latter forties with the Head of the Eco in the Antitrust Division of the Justice Department to di ments in the industry and to sound out the governmen price increase. As Lehmann puts it:

> In general the companies were given the impression tha
> not justified by a cost increase might appear (to the
> Division) as a monopolistic exploitation of consumers a
> investigation of the industry.[27]

Within a few months after the outbreak of war 1950, an impending shortage of sulphur became app ning of 1951, Frasch producers had to curtail shi mestic and export customers. In September 1950 F price by $4 per ton. Texas Gulf delayed until 30 then announced a price hike of $3 per ton. Thi difference in pricing between the two major pro lishment of Sulexco in 1922. Lehmann, basing h learned in personal contacts with company offi was motivated by a fear of antitrust prosecution

> The lagged differential pricing by Texas Gulf p
> rapidly developing world shortage, can be vie
> oriented. From conversations with Texas G
> state: (1) that the increase to $21 was m
> against government investigation and prosec
> some future time); (2) that the decision v
> legal counsel, who advised that $22 was a
> that the plan was to later raise the price to $

26. Glenn Albert Lehmann, "The Market f
(Ph.D. dissertation, Economics Department, H
11.
27. *Ibid.*
28. *Ibid.*, pp. 312–13.

in part, a response by the Frasch sulphur industry and others to market conditions during the shortage. While it is tempting to attribute all the change in the sulphur industry in the fifties to the sulphur shortage, it must be recognized that other developments, in particular the growth in production of sour natural gas, complemented the search for new sources of supply. Nevertheless, the influx of new capital into the sulphur industry in response to higher rates of return served to widen the base of the industry and brought changes in the pattern of market behavior which had remained unchanged for thirty years.

As mentioned above, the initial response to the sulphur shortage was the imposition of governmental controls over sulphur shipments and prices. At the same time, the government provided incentives, such as rapid tax amortization and exploration loans, to stimulate the development of new sulphur capacity. While domestic prices were controlled, the price on the open market, particularly in Europe, rose to very high levels with some sales at over $200 per ton being reported. In addition, some customers in Europe experienced long delays in obtaining their sulphur requirements. After the removal of price controls in March 1953, the price of Frasch sulphur rose in two successive increments to $26.50 per ton for domestic shipments f.o.b. the mine; $31.00–$33.00 per ton for export shipments f.o.b. vessels, Gulf port; and $28.00–$33.00 per ton for domestic and Canadian shipments f.o.b. vessels, Gulf port in January, and $28.00–$29.50 per ton from February to December.[31]

Entry of Mexican Frasch production

Production of elemental sulphur by the Frasch process in Mexico began in 1954, and reached nearly a half million tons in 1955. The Frasch producers in Mexico, led by the Pan American Sulphur Company, had initially intended to market their output mainly in Western Europe to take advantage of the $3.00 per ton differential between the export and domestic prices of U.S. Frasch producers.[32] Mexican Frasch sulphur was initially priced at $29.00 per ton f.o.b. the Mexican port, or $2.00 per ton below the comparable U.S. posted price. The policy proved to be so successful that, in February 1956, Freeport Sulphur cut its export price by $3.00 per ton in an attempt to maintain its share of foreign markets. In April, Texas Gulf reduced its export prices by $3.00 per ton, charging $28.00 per ton for bright sulphur to all destinations. Jefferson Lake and Duval made similar price reductions.

However, with the price umbrella removed from the European mar-

31. U.S. Department of the Interior, Bureau of Mines, *Minerals Yearbook*, 1954, vol. 1, p. 1127.
32. This information was confirmed in an interview with Mr. Henry Webb, President of Pan American Sulphur Company.

ket, Mexican Frasch sulphur producers decided that Mexican sulphur at $28.00 per ton f.o.b. the Mexican port was competitive in the United States as well as in other export markets. In addition, the Mexican firms had the advantage of being able to transport sulphur to the U.S. coastal markets at $1.50 per ton less than the U.S. Frasch producers who were required by law to use U.S. vessels to serve domestic ports.[33] Freeport, which had initiated the abolishment of the export differential and which sold most of its domestic sulphur in the inland markets of the Mississippi Valley, was not as much affected by the freight differential as was Texas Gulf, which supplied most of the East Coast markets.

By 1957, Mexican production of Frasch sulphur had reached nearly one million tons and had made marked inroads into both the domestic and export markets for U.S. Frasch sulphur as shown in figure 12. On 18 September 1957 Texas Gulf Sulphur Company announced a price reduction of $3 per ton on domestic and Canadian shipments of Frasch sulphur, thus reestablishing the export differential and bringing the domestic price to $23.50 per ton f.o.b. the mine, $25.00 per ton f.o.b. vessels, Gulf ports. Within hours, Freeport Sulphur Company announced a straight-across-the-board cut of $3 per ton on all shipments, both domestic and export.[34] Texas Gulf then amended its initial action, going along with Freeport's broader reductions in the export market. The other domestic producers followed with similar reductions.

Thus, Texas Gulf was unsuccessful in its attempt to reestablish the export differential and encourage Mexican tonnage to move to European markets. Two additional factors prompting the price reductions were the advancements made by Mexican producers in improving the grade of Mexican sulphur through filtering and a decision by Pan American Sulphur to build a stockpile at Tampa, Florida, thus increasing its competitive potential in the large Florida fertilizer market in which Texas Gulf was the largest seller.[35]

The 10 per cent reduction in Frasch sulphur prices made in 1957 was designed to enable domestic producers to regain the share of domestic and foreign markets which had been lost to Mexican Frasch sulphur. However, one Mexican producer, Gulf Sulphur Corporation, immediately reacted by reducing its posted price by $2.00 per ton to $23.00 f.o.b. the Mexican port for filtered (bright) Frasch sulphur,

---

33. W. G. Brese, *An Analysis of the Sulphur Industry in Alberta* (Calgary, Alberta: Research Council of Alberta, Information Series, No. 38, 1962), p. 44. The Mexican port of Coatzacoalcos was about equidistant with the major sulphur exporters of the Gulf Coast to the important sulphur markets on the Eastern Seaboard of the United States.

34. See *Chemical Week*, 28 September 1957, p. 85.

35. *Sulphur* (September 1957), p. 16.

while the other Mexican Frasch producer, Pan American Sulphur, reestablished its competitive position in the United States and overseas markets by adjusting discounts and freight allowances without resorting to changes in its posted prices. The price cutting did not succeed in stopping the entry of Mexican Frasch sulphur into the United States, though it may have been responsible for leveling off Mexican imports. As shown in figure 12, from 1957 through 1959, imports of Frasch sulphur from Mexico averaged about 600,000 tons per year. Over this period, the reduction in price, however, is estimated to have cost domestic Frasch producers $35 million in forgone revenue.[36]

1958 to 1963: years of surplus

Following the major price reduction in 1957, the posted price for Frasch sulphur became increasingly less meaningful. By 1958, all Frasch producers, both U.S. and Mexican, had begun quoting prices on a delivered basis, thus breaking the pattern of f.o.b. pricing adopted by Herman Frasch in the industry's infancy and maintained for over five decades. Freight contributions and discounts began to be used to obtain contracts in the more competitive markets, particularly along the East Coast of the United States.[37] In December 1960, Pan American Sulphur reacted to a brief upturn in sulphur markets by raising its posted price by $2.00 per ton. Domestic Frasch producers responded by withdrawing part of the transportation allowances which represented the gap between their posted price and actual realizations (see figure 13).[38]

The development of recovered sulphur production in Western Canada and at Lacq, France, brought additional supplies of elemental sulphur into world markets during the late fifties and early sixties and further reduced the market shares held by U.S. Frasch producers. The substitution of delivered pricing for f.o.b. pricing and the development of several sources of supply for elemental sulphur resulted in a trend toward regional pricing with relative freight rates becoming the decisive factor in securing markets.

The competitive position of the Canadian recovered sulphur industry in U.S. markets was greatly enhanced during 1961 by the reduction in

36. Brese, *op. cit.*, p. 54.
37. See *Sulphur* (September 1960), p. 4, where it was pointed out that posted prices "apply only in an ever decreasing and now very small number of protected markets and the bulk of domestic and export sales are based on 'delivered' prices which include freight contributions and discounts generally of up to $3.50 per ton although the recent intensification of the price war has augmented this concealed factor to over $5. In effect, therefore, the bulk of U.S. Frasch sulphur sells today at an f.o.b. equivalent of $21.50 for bright and $20.50 for off-colour materials, while some bright sulphur is sold at prices equivalent to as little as $18.50."
38. *Oil, Paint, and Drug Reporter* (26 December 1960), pp. 3 and 8.

rail freight rates from Alberta to Chicago from $19.94 per ton to
$12.88 per ton. The new rate, which became effective in August 1961
and which was unsuccessfully challenged in federal court by Freeport
Sulphur Company,[39] enabled Canadian recovered sulphur to be sold in
midwestern markets at about $26.00 per ton, a price that included
$12.00 to $13.00 in freight costs. Domestic Frasch producers responded
by reducing their delivered prices in markets facing competitive pressures from the Canadians. In 1963, both Texas Gulf and Freeport
challenged the rail rates from Alberta to midwestern markets in Inter-
state Commerce Commission hearings. The rail rate of $12.88 per ton
paid on shipments from Alberta to Chicago was substantially the same
as that paid by U.S. producers on shorter shipments from the Gulf
Coast to Chicago. Nevertheless, the challenge was denied on appeal in
January 1966.[40]

Domestic Frasch producers, after their futile attempts to maintain
their share of domestic and foreign markets by price competition with
the Mexican Frasch producers, adopted nonprice competition as a
secondary line of defense during the early sixties. The switch from
solid to liquid delivery of Frasch sulphur was particularly helpful in
maintaining domestic market shares. By agreeing to finance the conversion of customers' plants to the storage, handling, and use of liquid
sulphur, domestic producers were able to secure long-term contracts and
develop a closer tie between the customer and the supplier.[41] By placing
greater emphasis on customer service, the domestic Frasch producers
were able to retard the erosion of prices in domestic markets.

In export markets, the four domestic Frasch producers responded to
competition from Mexican, Canadian, and French sulphur by moving
in 1958 to reestablish the Sulphur Export Corporation, with the following distribution of ownership: Texas Gulf, 38 per cent; Freeport, 38
per cent; Jefferson Lake, 18 per cent; and Duval, 8 per cent. Sulexco
was reformed to bring unity to the efforts of domestic Frasch producers
in competing for foreign markets. In addition, Sulexco could offer its

39. U.S. Department of the Interior, Bureau of Mines, *Minerals Yearbook,*
1961, vol. 1, p. 17. Also see *Chemical Week* (6 January 1962), p. 21.

40. U.S. Department of the Interior, Bureau of Mines, *Mineral Industry Sur-*
veys (Sulphur), September 1963, p. 2. Also see Arthur D. Little, Inc., *The Free*
World Sulphur Outlook (April 1966), p. 17.

41. The leading trade journal made the following observation regarding the
effect of transition to liquid sulphur delivery: "... to date the liquid sulphur
programme has contributed significantly to the prevailing relatively stable marketing conditions in the U.S.A...." "Sulphur Price War—World Brimstone Prices
Drop Over $3," *Sulphur* (December 1962), p. 1. As with most customer-oriented
innovations, for example, the showing of in-flight movies on domestic airlines, the
switch to liquid sulphur delivery was soon implemented by competitors, in this
case, Pan American Sulphur Company, Gulf Sulphur Corporation, and SNPA.

customers technical aid and ocean freight assistance—matters of increasing importance with the growth of liquid sulphur deliveries.[42]

Some foreign consumers showed a marked aversion to purchasing from a marketing cartel. Differences in scale and nature of operations between the four members of Sulexco also hindered the development of a unified marketing policy. In particular, the willingness of Texas Gulf and Freeport to reduce prices in order to maintain tonnage appears to have worked to the detriment of Duval and Jefferson Lake, which were operating on marginal deposits with higher operating costs. As a result, they were not in a position to benefit from a sacrifice of incremental revenue for additional tonnage. Thus, Duval and Jefferson Lake were in the uncomfortable position of having only a minority interest in determining the policy under which 60 per cent of their output was being sold.[43]

Even with the formation of Sulexco, however, domestic Frasch producers continued to lose their share of world export markets. Between 1960 and 1963, as shown in figure 12, their share of world markets fell by 35 per cent. Not all of Sulexco's competition in world sulphur markets came from Mexican and Canadian sulphur. In 1963, SNPA, the state-controlled firm producing recovered sulphur at Lacq, France, undertook an aggressive, price-slicing policy to obtain a larger share of the world market and unload sizable inventories. It managed to increase its share of the Western European export market, particularly in the United Kingdom, by from 25 to 36 per cent. As a result, the proportion of U.S. sulphur exports to imports fell, and realizations on both domestic and foreign shipments hit a twenty-year low (see figure 13).

1964 to 1967: years of shortage

The sulphur market abruptly reversed itself in mid-1963, and suppliers and consumers found themselves faced with a shortage for which they were unprepared. The underlying forces causing this switch were the restricted growth in sulphur capacity during the early sixties when prices were depressed, and a rapid surge in demand, both at home and abroad, particularly in the chemical fertilizer industry. From 1963 through 1967, sulphur consumption continued to rise, increasing 38 per cent in the United States and 32 per cent in the noncommunist countries. While production of sulphur in all forms increased 29 per cent, it could not keep pace with demand. The result was a drawdown in sulphur stocks of about 1 million tons in 1964, 1.1 million tons in

---

42. Brese (*op. cit.*, p. 48) estimated that Sulexco was able to reduce handling and shipping charges by as much as $3.00 per ton through pooling its shipments and transporting sulphur in large quantities.
43. *Chemical Week* (11 June 1960), p. 101.

1965, 0.6 million tons in 1966, and 0.3 million tons in 1967 (see figure 11).

Reacting to the pressure for increased supply, Texas Gulf initiated a $2.00 per ton increase in the posted price of sulphur, effective 1 July 1964. Other U.S. producers and Mexican producers quickly followed Texas Gulf's lead. However, discounts by domestic Frasch producers remained at $1.50 per ton where Mexican sulphur was sold and $3.00 per ton where the competition was Canadian.[44] Throughout 1965, published prices for domestic sales remained unchanged at $27.00 per ton f.o.b. Gulf ports and $25.50 per ton f.o.b. mines, with $1.00 per ton less for dark or acid grade sulphur. However, as shown in figure 13, producers gradually terminated competitive discounts and began a program of charging realistic delivered prices that fully covered transportation and terminal costs.

Undoubtedly, the domestic posted price for sulphur would have been raised during 1965 had not the President's Council of Economic Advisors and other administrative agencies made it clear that such action would bring about a quick response from a highly price-conscious administration.[45] During 1965, the export price for bright Frasch sulphur climbed from $27.50 per ton in January to $31.00 per ton on 15 February and to $36.00 per ton in mid-June. The widening differential between domestic and foreign prices during 1965 made exports extremely attractive to domestic Frasch producers (see figure 13), and they responded, as shown in figure 12, by increasing their exports to 2.67 million tons, an all-time high representing 32.4 per cent of total U.S. production in 1965.

During 1965, it became apparent that further expansion of exports would lead to shortages in the domestic market and increase pressures on the domestic price. Thus, for Frasch producers, it became a question of selling abroad at higher prices thereby creating a domestic shortage and threatening the posted price, or of curtailing exports in order to satisfy domestic demands at the existing posted price. Government pressures eventually forced Frasch producers to restrict their exports. During 1966, Sulexco worked with the State Department in an informal arrangement to ration sulphur exports. Sulexco sent to the State Department an estimate of exports for 1966 broken down by country. The Department in turn informed Sulexco of its "views" regarding total

---

44. *Oil, Paint, and Drug Reporter* (6 July 1964), p. 27.

45. While not generally publicized, Administration efforts to persuade domestic producers to hold the line on sulphur prices were quite effective. During 1965, the Administration, acting through the Council of Economic Advisors, turned down requests by the major domestic Frasch producers to increase their domestic prices. Had sulphur prices risen without approval by the Administration, it is likely that the industry would have faced either export controls or antitrust action.

export tonnage and its allocation among countries.[46] The rationing of exports produced difficulties in some markets that were hard hit by a reduction in exports from Mexico due to a production failure on a portion of Pan American's mine.

Early in 1966, the two Mexican Frasch producers increased their price for sulphur to U.S. consumers by $5.00 per ton. The major U.S. Frasch producers did not respond by increasing their domestic prices, but continued the program begun in 1965 of eliminating allowances and adjusting transportation and handling charges in order to recover the costs of transportation and terminals. Realizations on domestic sales, as shown in figure 13, climbed throughout 1965 and 1966. Sulphur markets continued to tighten during 1966, bringing additional pressures on the domestic posted price. Finally, effective 1 December 1966, Freeport Sulphur increased its domestic price by $2.50 per ton for dark sulphur f.o.b. Port Sulphur, the company's main shipping point. Transportation and handling charges for sulphur shipped by Freeport beyond Port Sulphur remained unchanged from the levels established earlier in 1966. Texas Gulf increased its price for domestic sales by $2.50 per ton to $28.00 per ton for bright sulphur f.o.b. mines, effective 15 December 1966.[47]

In the middle of December, Pan American announced an increase of $10.00 per ton on all sulphur sold in the United States effective 1 January 1967. Gulf Sulphur reported that upward price adjustments had been made in its contracts as they came up for renewal. Prices in other parts of the world continued to run substantially above those in the domestic market. Toward the end of 1966, export prices for U.S. Frasch sulphur were $39.00 per ton and upwards for bright sulphur f.o.b. U.S. Gulf ports (see figure 13). Export prices for Mexican Frasch sulphur and recovered sulphur were reported to be at even higher levels.

The continuing sulphur shortage led to two price increases for domestic sulphur in 1967. Published prices increased $4.00 per ton f.o.b. Gulf ports in April and were increased an additional $5.50 per ton in October. Year-end prices for bright sulphur were $39.00 per ton f.o.b. Gulf ports. Like the 1966 increase, the 1967 price hikes were initiated by Freeport, indicating its position as the leading Frasch producer. The increases in posted prices during 1966 and 1967 were insufficient to ration existing supplies of sulphur among consumers. It became necessary for sulphur deliveries to be allocated by the major producers. In

46. This arrangement was described by Edward Getzin, Chief, Industrial and Strategic Minerals Division, United States Department of State, in an interview with the author in Washington, D.C., on 11 August 1966.

47. L. B. Gittinger, Jr., "Sulphur," *The Engineering and Mining Journal* (February 1967), p. 170.

September 1966, Texas Gulf cut deliveries to domestic consumers to 75 per cent of their 1965 level, and in December 1967, customers were notified that they would be limited in 1968 to 65 per cent of their 1965 purchases.[48] Freeport, early in 1967, told customers it could deliver only 90 per cent of their base tonnage.

Export prices continued to rise during 1967. In August, Pan American Sulphur increased its run-of-the-mill sulphur price to $50.00 per ton for sales to regular contract customers and to $55.00 per ton for spot sales, f.o.b. Coatzacoalcos, Mexico. Bright sulphur prices were increased to $52.00 and $57.00 per ton respectively, maintaining the $2.00 per ton difference.[49] Canadian recovered sulphur, early in 1968, was reported to be selling as high as $56.50 per ton f.o.b. Vancouver.[50]

1968: balance achieved

World sulphur markets returned to balance in 1968 when, for the first time since 1962, production exceeded shipments in the noncommunist countries. The return to balance between supply and demand was attributed to two factors: (1) an increase in sulphur production, primarily brimstone production in Western Canada augmented by increased output of Frasch sulphur in the United States and Mexico; and (2) a reduction in the rate of growth of sulphur consumption, primarily due to the fact that there was no increase in U.S. consumption.[51] Late in 1968, prices in Western Europe and some other overseas markets were reduced by $2.00 per ton. In January 1969, similar price reductions were made in the domestic and Canadian markets. With the sulphur shortage at an end, both suppliers and consumers were left pondering whether the feast or famine pattern of sulphur markets during the past two decades would be repeated in the coming years. This question is considered in chapter 8.

---

48. "Texas Gulf Again Cuts Sulphur Allocations to 65% of '65 Deliveries, as Shortage Grows," *The Wall Street Journal* (5 December 1967), p. 34.

49. "Mexican Concern Increases Prices On Its Sulphur," *The Wall Street Journal* (29 August 1967), p. 28.

50. Thomas O'Hanlan, "The Great Sulphur Rush," *Fortune* (March 1968), p. 109.

51. L. B. Gittinger, "Sulphur," *The Engineering and Mining Journal* (March 1969), pp. 160C–160F.

Part III

Performance in the Sulphur Industry

Chapter 7

Evaluation of Competitive Performance

Analysis of the market structure and behavior of an industry provides observations upon which an evaluation of the industry's performance can be made. However, in contrast to the analysis of market structure and behavior, performance evaluation is largely qualitative in nature. It is simply a judgment of an industry's effectiveness vis-à-vis some norm or standard. The commonly accepted norm is that of pure competition. However, the use of the competitive standard is based upon an implicit value judgment. Given different value judgments, other norms might replace the competitive standard. For example, if progressiveness is an accepted criterion of performance, the competitive standard may not offer any basis for comparison.[1] If the problem of resource utilization is a matter of sole concern, monopoly may be preferred to competition.[2] Even if one accepts the value judgments behind the use of the competitive standard, its application to any given industry is very difficult. In the final analysis, evaluation of the performance of an industry must rest upon an appraisal of the overall impact of industry behavior upon what is thought to represent the public welfare.

Earlier chapters in this study pointed up the oligopolistic nature of the Frasch sulphur industry's market structure. Analysis of the industry's market behavior revealed a rigid pricing policy and the existence of price discrimination between domestic and export markets over much of the industry's history. However, it was also revealed that the indus-

1. Phillips, *Competition in the Synthetic Rubber Industry*, p. 202, states ". . . there are no analytical standards to measure progressiveness since innovation has not been dealt with by competitive models." The same point is made by Hamilton, *Competition in Oil*, p. 153.

2. Anthony Scott, *Natural Resources: The Economics of Conservation* (Toronto: The University of Toronto Press, 1955), p. 84, contends that monopoly, through the restriction of output, serves the cause of resource conservation more effectively than competition. "It may even follow that when conservation policy has the higher priority, the state should actively encourage monopoly as an ally in this programme."

111

try's market behavior has undergone a significant change during the past decade in response to fundamental alterations in its market structure. Evaluation of the industry's performance, therefore, must be made not only in light of the industry's history, but also in light of the recent changes in market structure and behavior and their implications for future performance.

The most important indicators of industry performance are the level of profits and the relationship between costs and prices. These factors are examined in detail in this chapter. Another indicator—and one that is important in a resource-based industry—is the utilization of the sulphur resource and the outlook for sulphur as a mineral resource. This aspect of performance is examined in the following chapter.

The Level of Profits

Profit levels as a competitive standard

The clearest indication of how well an industry meets competitive standards is the long-run record of its profits. A basic premise of competition is that the rate of return to capital invested in various sectors of the economy, for a given degree of risk, will be approximately equal, given the time required for adjustment. Thus, while an industry may experience an above-normal rate of return during periods of rapidly growing demand or dynamic innovation, these profits will in time be bid down to competitive levels through the attraction of new capital into the industry, the expansion of output, and the reduction of prices. While the process of equalization may be retarded by factors limiting the mobility of capital, particularly at the international level, evidence indicates that the economy in general performs this task fairly well.[3] Of course, it should be emphasized that while persistent above-average profits are almost *prima facie* evidence of noncompetitive performance, the existence of an average rate of return in an industry does not prove that the industry is competitive.[4] However, as a first approximation of competitiveness, the level of profits over the long run is a very useful tool for evaluating the performance of an industry.

Profits in the Frasch sulphur industry

The Frasch sulphur industry developed from an ingenious and unique innovation, the Frasch hot-water mining process. It would be expected

3. George J. Stigler, *Capital and Rates of Return in Manufacturing Industries* (Princeton: Princeton University Press for the National Bureau of Economic Research, 1963).
4. *Ibid.*, p. 55.

Be that as it may, the decision proved to be an extremely costly one to Texas Gulf. In 1951, the Office of Price Stabilization, under the provisions of the General Ceiling Price Regulation, froze sulphur prices at levels prevailing during the base period, 19 December 1950 to 25 January 1951. As it was not until 17 March 1953 that controls were removed on Frasch prices, Texas Gulf's price decision may have cost the company in the neighborhood of $3 million in forgone revenue.

During the Korean crisis controls were placed on domestic usage and exports, as well as on prices. Thus, there was little opportunity for the sulphur shortage to affect the behavior of Frasch producers at this time. The sulphur shortage is given credit, however, for the dissolution of Sulexco in 1952. Freeport requested the dissolution of Sulexco. In an interview, one industry official stated that, with sulphur being rationed by the government, there was little point in domestic producers continuing the expense of maintaining a sales organization for export shipments. At the time, there was certainly no need to encourage sulphur sales abroad; the problem was to limit export demands so that domestic demands could be met.

New sources of sulphur emerge

In June 1952, the President's Materials Policy Commission, established by President Truman, published *Resources for Freedom,* commonly called the "Paley Report," which presented a comprehensive review of the supply and demand for raw materials in the United States and the free world, including projections to the decade 1970–80.[29] The commission projected a 110 per cent increase in the demand for sulphur by 1975, and concluded that the country could not depend on major increases in the supply of Frasch sulphur to meet this expansion in demand. The Commission urged increased effort to recover more sulphur from pyrites and sour gases.

A decade later, in his review of the portion of the Paley Report pertaining to sulphur, William L. Swagger concluded: "Technology has been successful in providing increasing supplies of sulphur at no increase in real costs as the Paley Commission warned it would be required to do."[30] Between 1953 and 1960, aided by discoveries of additional deposits and technological improvements in recovery techniques, new sources of supply of elemental sulphur were developed both within the United States and abroad.

The development of new sources of supply in the years immediately following the sulphur shortage of the early fifties represented, at least

29. The President's Materials Policy Commission, *Resources for Freedom* (Washington, D.C.: U.S. Government Printing Office, 1952).
30. William L. Swagger, "The Paley Report in Review: Sulphur," Battelle Memorial Institute, 1961, p. 13.

1,366,361 tons between 1947 and 1950. Responding to the expansion in demand, Freeport raised its price from $16 to $18 per ton f.o.b. the mine in June 1947, and two months later Texas Gulf met the price increase (see figure 13).

Glenn Lehmann, in his study of the industry, has emphasized the effective role played by the Justice Department in the immediate post-war period in preventing additional price increases on Frasch sulphur.[26] Industry officials had experienced public scrutiny of their operations by the Temporary National Economic Committee of Congress in 1939 and by the Federal Trade Commission in 1947, and they were anxious to avoid any action that might draw public attention to the concentrated nature of the industry. Representatives of the major producers met frequently during the latter forties with the Head of the Economic Section in the Antitrust Division of the Justice Department to discuss developments in the industry and to sound out the government reaction to a price increase. As Lehmann puts it:

> In general the companies were given the impression that a price increase not justified by a cost increase might appear (to the public or to the Division) as a monopolistic exploitation of consumers and thus lead to an investigation of the industry.[27]

Within a few months after the outbreak of war in Korea in June 1950, an impending shortage of sulphur became apparent. By the beginning of 1951, Frasch producers had to curtail shipments to both domestic and export customers. In September 1950 Freeport increased its price by $4 per ton. Texas Gulf delayed until 30 November 1950, and then announced a price hike of $3 per ton. This represented the first difference in pricing between the two major producers since the establishment of Sulexco in 1922. Lehmann, basing his judgment on what he learned in personal contacts with company officers, suggests the action was motivated by a fear of antitrust prosecution:

> The lagged differential pricing by Texas Gulf particularly in the midst of a rapidly developing world shortage, can be viewed as entirely government oriented. From conversations with Texas Gulf officials, this writer can state: (1) that the increase to $21 was made as an added precaution against government investigation and prosecution (either at that time or some future time); (2) that the decision went counter to the advice of legal counsel, who advised that $22 was a perfectly safe price; and (3) that the plan was to later raise the price to $22.[28]

26. Glenn Albert Lehmann, "The Market for Sulphur: A Study in Duopoly" (Ph.D. dissertation, Economics Department, Harvard University, 1953), pp. 310–11.

27. *Ibid.*

28. *Ibid.*, pp. 312–13.

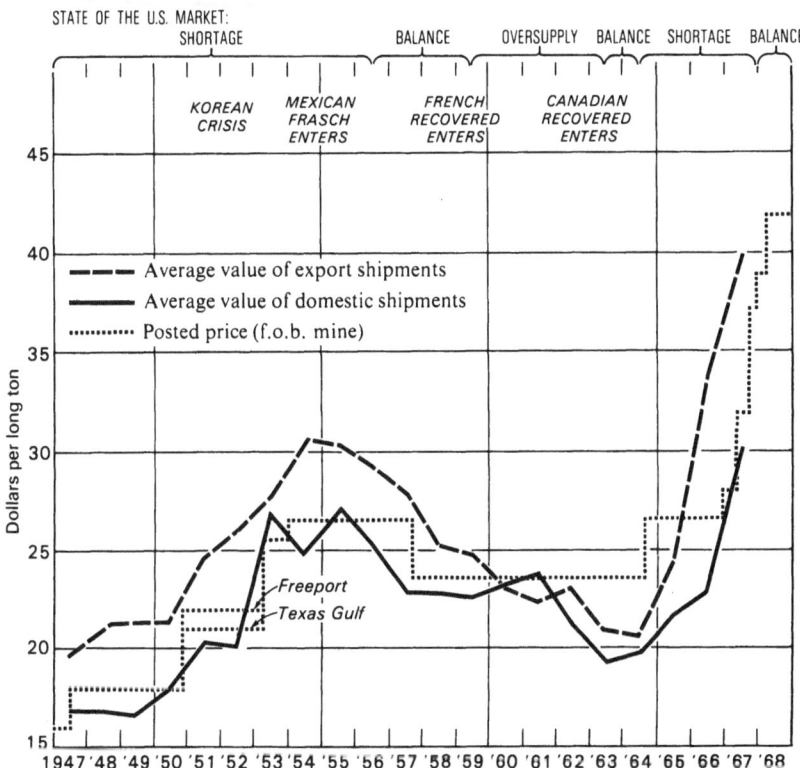

Figure 13. Average value of export and domestic shipments of U.S. Frasch sulphur and posted price, f.o.b. mine, 1947–68. (Based on data in tables 15 and A-3.)

agreement or understanding with nonmember domestic producers; and from selling or assisting in the sale of output from nonmember domestic producers.[25] However, by 1947 these were more or less academic issues because the Frasch sulphur industry had divorced itself from its international agreements during the war, and the Sulexco members had made no further attempt to obtain agreements with either domestic or foreign producers.

As shown in figure 11, the demand for Frasch sulphur increased rapidly in the years following the end of World War II. The largest domestic consumer of sulphur, the sulphuric acid industry, increased its output from 766,800 short tons in 1946 to 1,085,000 short tons in 1950. Exports of crude sulphur, depicted in figure 12, which had averaged 566,361 tons between 1935 and 1939, increased to an average of

25. *Ibid.*, pp. 102–3.

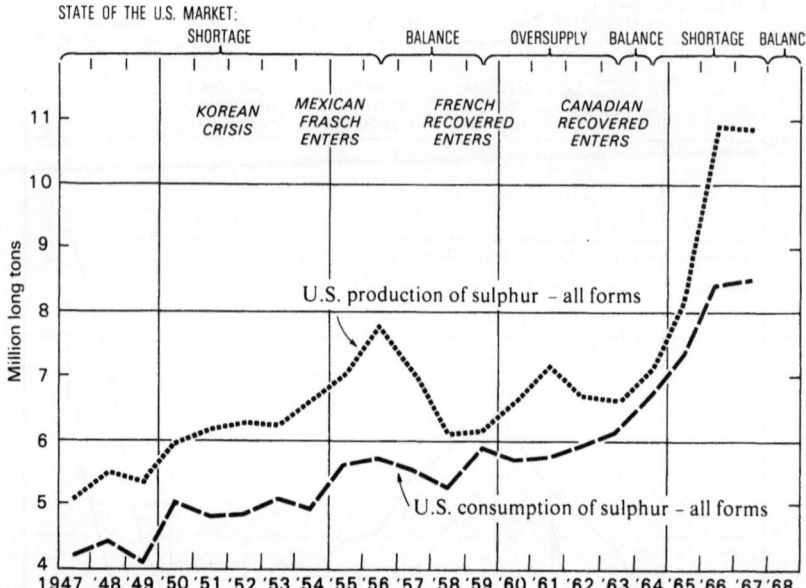

Figure 11. U.S. sulphur supply and demand cycle, 1947–67. (Based on data in tables A-1 and A-2.)

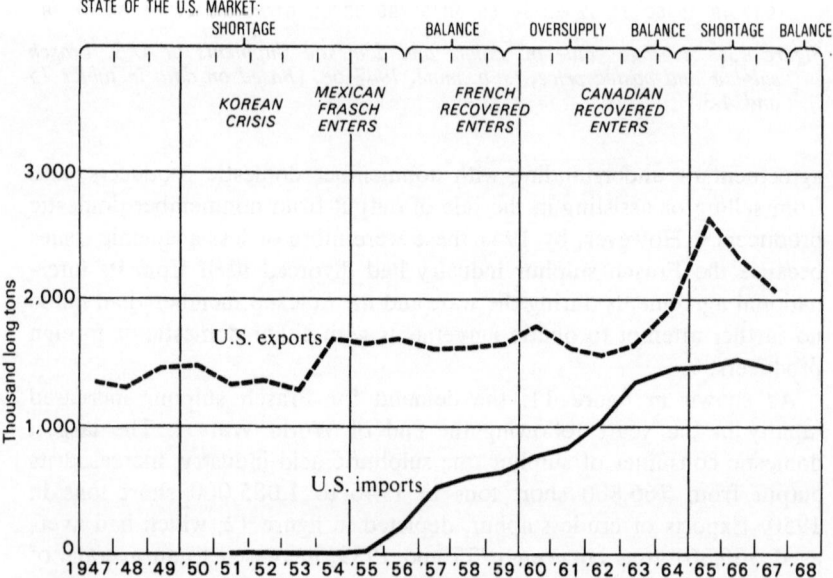

Figure 12. U.S. exports and imports of all forms of sulphur, 1947–67. (Based on data in table A-2.)

that after perfecting the process, the innovator would have received a rate of return during the industry's early phase that would have been significantly above the normal rate of return for capital invested in other parts of the economy. The existence of above-normal rates of return, however, should have induced entry into the industry and lowered the rate of return. Unfortunately, data on profits earned by the Union Sulphur Company are not available. In any event, new producers were soon attracted into the industry.

In its 1947 study of the sulphur industry, the Federal Trade Commission computed average invested capital, profits before and after taxes, and the rate of return before and after taxes for the four principal Frasch sulphur producers over the period from 1919 to 1946. During this period, Texas Gulf, Jefferson Lake, and Duval were engaged only in the production of Frasch sulphur. Freeport was also engaged in mining manganese and nickeliferous iron ore in Cuba during this period, but the Commission excluded these operations so that the reported rate of return on investment represented only profits on the production and sale of sulphur. To compensate for the high income and excess profits taxes during the period from 1941 to 1946, the Commission calculated the before-tax rate of return by relating the profit before federal taxes to the average invested capital adjusted for such tax provisions.

Since 1947, data on profits earned on Frasch mining operations are less reliable because all Frasch producers have diversified their operations into related fields of mining and chemicals. Accounting data published in the annual reports of these firms incorporate all their activities, and separate information regarding investment in Frasch mining and profits earned on Frasch sales is not available. Nevertheless, since sulphur mining remained the principal activity of both Texas Gulf and Freeport through 1966 and Jefferson Lake through 1962 (in 1964, Jefferson Lake was acquired by Occidental Petroleum), accounting data for these firms provide a reasonably good basis for computation of the rate of return on Frasch mining activities. However, over the past two decades, sulphur mining has become an increasingly smaller element of Duval's operations and has not been a dominant activity of the firm since 1953. Thus, table 17, which lists average invested capital, after-tax profits, and the after-tax rate of return on invested capital both for the entire Frasch industry and for each of the individual producers, includes no data for Duval after 1953 and no data for Jefferson Lake after 1962.

Table 17 shows that the average rate of return on invested capital for all firms in the Frasch sulphur industry was 23.60 per cent for 1919–53 and 21.23 per cent for 1919–66. Most economists would agree that

Table 17.   Average Investment and After-Tax Profits and Rate of Return
for the Frasch Sulphur Industry, 1919–66

	Rate of return after taxes on invested capital				Frasch Sulphur Industry		
Year	Texas Gulf	Freeport	Jefferson Lake	Duval	Invested capital*	Income after taxes	Rate of return
	per cent	per cent	per cent	per cent	thousand dollars	thousand dollars	per cent
1919	18.76	11.48	—	—	14,791	2,073	14.02
1920	37.04	6.94	—	—	17,387	3,911	22.49
1921	16.58	(5.82)	—	—	19,727	1,374	6.97
1922	31.24	(2.53)	—	—	22,198	3,555	16.02
1923	36.39	6.33	—	—	25,042	5,453	21.77
1924	36.01	(2.65)	—	—	25,502	4,437	17.40
1925	42.44	6.37	—	—	25,032	6,378	25.48
1926	62.85	14.40	—	—	27,356	11,105	40.59
1927	67.86	27.01	—	—	31,516	15,738	49.94
1928	66.49	24.59	—	—	34,991	17,871	51.07
1929	58.66	33.07	—	—	39,210	20,028	51.08
1930	41.99	23.94	—	20.66	44,826	16,706	37.27
1931	24.99	18.51	—	14.76	47,496	11,059	23.28
1932	16.34	21.35	—	9.98	47,089	8,067	17.13
1933	19.85	22.94	35.77	(6.54)	51,971	10,687	20.55
1934	13.69	11.70	33.85	3.61	65,501	9,124	13.93
1935	12.50	12.07	5.51	6.54	75,686	9,170	12.12
1936	16.66	16.73	n.a.	24.03	73,565†	12,140†	16.50†
1937	19.60	17.30	(0.51)	14.07	77,011	14,242	18.49
1938	11.75	10.87	33.27	19.69	77,658	9,672	12.45
1939	13.22	10.87	47.01	26.92	79,516	11,447	14.40
1940	15.40	12.36	28.70	20.62	82,556	12,701	15.38
1941	15.34	13.63	10.46	25.13	83,198	12,585	15.13
1942	14.97	10.60	(13.81)	28.99	83,413	11,329	13.58
1943	13.55	12.54	6.46	30.91	82,887	11,338	13.68
1944	16.38	11.57	16.20	31.12	82,582	12,984	15.72
1945	16.91	14.40	13.40	30.00	84,546	13,667	16.17
1946	24.79	15.52	11.52	35.73	89,204	19,987	22.30
1947	32.86	9.10	11.41	25.65	103,396	25,843	24.99
1948	42.11	12.77	11.55	19.45	97,135	29,503	30.37
1949	47.47	17.01	13.61	21.64	92,323	31,051	33.63
1950	45.13	19.32	28.65	18.28	104,340	35,311	33.84
1951	39.44	16.26	26.13	14.05	117,459	34,365	29.26
1952	36.48	17.82	21.67	25.60	126,658	36,451	28.78
1953	33.37	19.43	23.58	24.78	124,638	37,772	30.31
1954	36.56	20.95	30.22	—	139,515	42,946	30.78
1955	33.75	21.82	21.03	—	161,751	46,666	28.85
1956	26.51	19.94	18.52	—	182,442	43,237	23.70
1957	15.92	17.88	13.24	—	193,037	31,924	16.54
1958	11.95	12.99	(6.05)	—	235,420	27,341	11.61
1959	11.58	10.11	(11.74)	—	278,199	27,341	9.83
1960	10.63	9.47	1.68	—	277,610	25,464	9.17
1961	10.09	9.19	10.15	—	277,505	26,724	9.63
1962	9.17	8.75	10.76	—	288,879	26,031	9.01
1963	6.58	8.43	—	—	295,111	22,248	7.54
1964	7.63	9.55	— ¹	—	312,872	26,967	8.62
1965	11.02	12.71	—	—	337,325	40,084	11.88
1966	14.11	17.17	—	—	389,869	60,844	15.61
1919–53 Avg.	—	—	—	—	—	—	23.60
1919–66 Avg.	26.35	13.68	15.59	20.28	—	—	21.23

SOURCE: Appendix B.
NOTE: Jefferson Lake was acquired by Occidental Petroleum in 1964; data not available after 1962. Data for Duval not available after 1953; owing to diversification, accounting data no longer reflect results from sulphur mining.
( ) Loss.   * Beginning and end of year.
n.a.—Not available.   † Excludes Jefferson Lake.

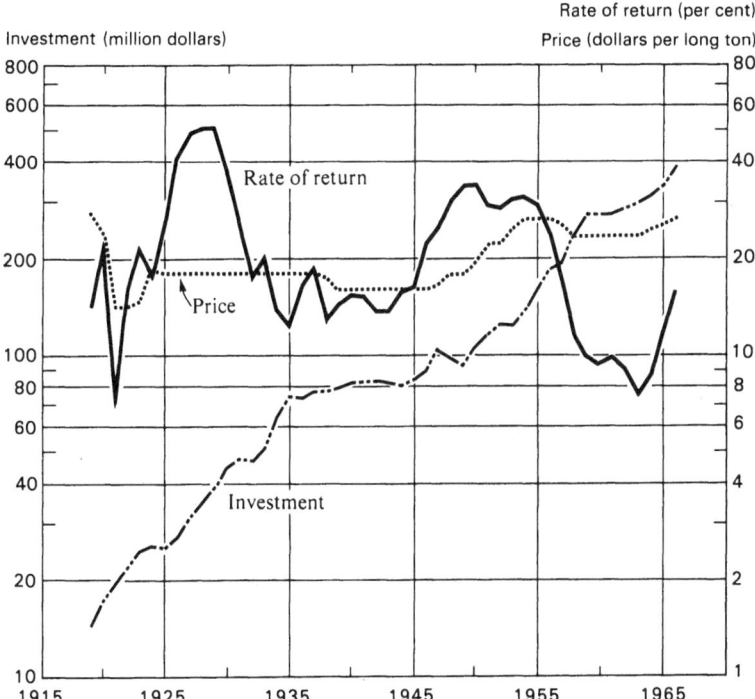

Figure 14. Annual average investment, price, and rate of return for the U.S. Frasch sulphur industry, 1919–66. (Based on data in tables 17 and A-3.)

these are abnormally high rates of return for any industry over such a long period.[5] In figure 14, the rate of return for the industry, the average annual investment, and the posted price for Frasch sulphur are plotted for the period 1919 through 1966. Under competitive conditions, it would be expected that as investment grew both the rate of return and the price would decline. However, as figure 14 shows, while investment increased steadily in the Frasch sulphur industry, the additional capital came not from new entrants but from expansion by existing producers, and the price of sulphur has shown no tendency to decline. The rate of return for the industry, except for the period from 1958 through 1965, remained above 12 per cent.

An exceptionally high level of profits was earned during the period

5. For example, John G. McLean and Robert W. Haigh, *The Growth of Integrated Oil Companies* (Boston: Division of Research, Graduate School of Business Administration, Harvard University, 1954), pp. 680–81, report that the average rate of return on invested capital for 120 oil companies between 1920 and 1952 ranged from 6.6 per cent to 10.9 per cent. Phillips, *op. cit.*, p. 210, found synthetic rubber firms earned from 6 per cent to 9 per cent on their investment.

from the formation of the Sulphur Export Corporation in 1922 to the commencement of the depression in 1930. The average return for all firms during this period was 38.82 per cent. While the depression reduced industry profits, the average rate of return for all firms in the industry between 1930 and 1939 was still a very healthy 17.54 per cent. Profits in the industry remained at a high level until about 1953 when they began to fall. The decline in profits continued unabated through 1963. Thus, the pattern of profits, like those for prices and output, changed significantly in the mid-fifties, responding to the development of alternative sources of supply for elemental sulphur, particularly in Mexico, Canada, and France, which produced major changes in the market structure and behavior of the domestic Frasch industry.

Table 17 shows considerable divergence between the rates of return earned by the individual firms producing Frasch sulphur. Texas Gulf, largely on account of its ownership of the industry's lowest cost and largest deposit, Boling Dome, led in profits, earning an average of 26.35 per cent on invested capital over the period from 1919 through 1966. The two smaller firms, Duval and Jefferson Lake, averaged 20.28 per cent and 15.59 per cent respectively over the periods shown in table 17. Freeport Sulphur averaged only 13.68 per cent for 1919–66. Comparison of the profits earned by Texas Gulf and Freeport reveals that Freeport earned considerably lower profits in the twenties and in the period from 1940 to 1957. This can be attributed to Freeport's having higher costs because it was operating on domes of lower quality than Texas Gulf during these periods. In particular, Freeport's Grande Ecaille mine was located in the Louisiana swamps where the costs of producing sulphur were higher than in the salt plains of the Texas coast. Since 1957, profits of both Freeport and Texas Gulf have followed a similar pattern, although Freeport has had a slightly higher rate of return. Freeport has also become the leading producer of Frasch sulphur during this period.

Table 18 compares the rates of return on invested capital after taxes for the domestic Frasch sulphur producers with the rates of return earned by the two Mexican producers. (Duval is omitted from the comparison because sulphur constituted a minor part of its operations over much of this period.) It can be seen that Pan American consistently exceeded the rate of return of the major domestic producers throughout this period. The return earned by Gulf Sulphur, after its initial period of losses, also exceeded the comparative rates of return earned by the domestic Frasch producers.[6]

6. PASCO's mine is on a dome of high quality and its costs are undoubtedly similar to those of the major U.S. Frasch mines (Boling Dome and Grande Ecaille). However, Gulf Sulphur's mine is of average quality and its costs are

*Table 18. Frasch Sulphur Producers: Comparative Rates of Return on
Invested Capital After Deduction of Income Taxes, 1955–66*

Year	Pan American	Gulf Sulphur*	Texas Gulf	Freeport	Jefferson Lake†
1955	6.60		33.75	21.82	21.03
1956	29.33		26.51	19.94	18.52
1957	34.09		15.92	17.88	13.24
1958	27.11		11.95	12.99	(6.05)
1959	20.97		11.58	10.11	(11.74)
1960	16.28		10.63	9.47	1.68
1961	12.30	4.94	10.09	9.19	10.15
1962	15.36	22.75	9.17	8.75	10.76
1963	16.48	15.06	6.58	8.43	
1964	22.02	10.99	7.63	9.55	
1965	17.03	20.66	11.02	12.71	
1966	28.85	22.66	14.11	17.17	
Average	20.54	16.18	14.08	13.17	7.20

SOURCE: Appendix B.
( ) Loss.
* No operating profit prior to 1961.
† Data not available after 1962; acquired by Occidental Petroleum, 1964.

Several limitations on the use of accounting data to obtain rates of
return should be mentioned at this point. The accounting concepts used
in calculating corporate profits differ in many respects from the corresponding economic concepts appropriate to the measurement of income.[7]
Similarly, the use of invested capital probably does not reflect the actual
value of the firm's investment in plant and equipment. While these
limitations are important and must be recognized, the use of accounting
data for calculating rates of return on investment must rest on the
practical grounds of data availability.

Depletion and royalties

Two additional aspects of using accounting data for computation of
rates of return on invested capital for the Frasch sulphur industry should

---

more nearly in line with the smaller domestic mines. The $1.50 per ton freight
differential on shipments to U.S. buyers probably helps to explain the higher
profits earned by the Mexican firms. See Dale B. Truett, "Sulphur and the Development of a Chemical Fertilizer Industry in Mexico" (Ph.D. dissertation, The
University of Texas, 1967).

7. For a comprehensive discussion of this problem, see Edgar O. Edwards and
Philip W. Bell, *The Theory and Measurement of Business Income* (Berkeley: University of California Press, 1961). The major problems are accounting for capital
gains and losses as they arise; distinguishing between gains from normal operations
and gains from merely holding assets as prices rise; and promptly recognizing
changes in the general price level which affect both the level of reported profit
and the statement of net worth.

be noted. Percentage depletion increases the after-tax earnings of Frasch sulphur firms, while royalties, which are taken as a cost of doing business, reduce the income of Frasch sulphur firms.

In 1932, sulphur was placed under the depletion provisions, which allow depletion of up to 23 per cent of gross income from the sulphur-producing property but not to exceed 50 per cent of net income. The extent to which the net income limitation restricts the percentage depletion claimed is a function of the relationship between net and gross income and the statutory rate of depletion. In the case of sulphur, Frasch producers are affected by the net income limitation if their ratio of net to gross income is less than 46 per cent.[8] However, data indicate that the net income limit does not severely restrict the amount of depletion claimed by Frasch producers. Depletion claimed by these producers was equal to 20.3 per cent in 1958, 20.6 per cent in 1959, and 24.2 per cent in 1960 of gross income from mineral extraction.[9]

The available evidence suggests that the main result of depletion in the Frasch sulphur industry has been to increase profits, rather than to stimulate exploration and development of new sulphur deposits. This is understandable since sulphur is one mineral which is seldom found intentionally. Indeed, most of the existing Frasch sulphur deposits were discovered in the process of petroleum exploration. It is not surprising, then, that exploration expenditures in the sulphur industry have been extremely low related to the average net income of firms in the industry or to the amount of percentage depletion which the industry receives.[10] In recent years, exploration expenditures have increased, but exploration efforts have been largely unsuccessful. In one of the few major

---

8. U.S. Congress, House Committee on Ways and Means, *President's 1963 Tax Message*, Hearings on the tax recommendations of the President contained in his message to the Congress, 24 January 1963, pt. 1, 88th Cong., 1st Sess., 1963, p. 297.

9. *Ibid.*, p. 309. Total depletion claimed can exceed both the gross rate of percentage depletion and the 50 per cent unconsolidated net income from the taxpayer's mineral activities, because: (1) percentage depletion must be computed on a property-by-property basis, and cost depletion must be used where higher than percentage depletion; (2) in deriving total net income, losses from producing properties and deductions on nonproducing properties are subtracted from positive net income.

10. For example, see James R. Nelson, "Percentage Depletion and National Security," in *Federal Tax Policy for Economic Growth and Stability*, Joint Committee Print, of the 84th Cong., 1st Sess. (Washington, D.C.: United States Government Printing Office, 1955), p. 473. In describing depletion in the sulphur industry, Nelson states: "During the 19 years 1930 through 1948, average net income was $13 million per year; average dividend payments, $11 million; and average annual exploration expense, $369,603. In 1946, allowable depletion was $11.2 million, and in 1947, $13.8 million. Total exploration expense was $7 million during the entire period, 1930–1948." Due to the diversification of sulphur firms into other mining activities during the fifties, these figures are the latest available on sulphur exploration expenditures.

explicit efforts to find Frasch sulphur, the off-shore exploration program of the major Frasch companies in the mid-fifties, and again in the mid-sixties, the expenditure of sizable funds failed to turn up any new commercial deposits of Frasch sulphur.[11]

In addition to percentage depletion, royalties also play an important role in determining the rate of return on investment in the Frasch sulphur industry. The rates of return shown in tables 17 and 18 were calculated after deductions for royalty payments had been made. In general, two types of royalty payments are made on sulphur deposits (see table 19). The first type—a fixed fee per ton of sulphur produced—is paid to the owner of the mineral rights to the property. A second

Table 19.   Summary of Royalties Paid by Frasch Producers

Company and mine	Description of royalties
Freeport:	
Hoskins	Subleased from The Texas Co. A fixed royalty of $1.06 per ton is payable to the fee owner, and 70 per cent of the net profits after deducting the fixed royalty is payable to the Texas Co.
Grande Ecaille	Subleased from Gulf Oil Corp., Humble Oil and Refining Co., and Shell Canadian Exploration Co. Fixed royalties to owners are $1.05 per ton. An additional fixed royalty of $1.00 per ton is payable to the three oil companies. The three oil companies also receive an additional royalty based on the sales price for sulphur. Freeport's share of the net profits from this property amounts to between 60 and 65 per cent of the total.
Lake Pelto, Bay Ste. Elaine, and Garden Island Bay	Subleased from The Texas Co. Net profits are shared equally with The Texas Co.
Grand Isle	Leased from Humble Oil and Refining Co. Net profits after recovery of all construction and development costs are shared equally with Humble.
Chacahoula, Nash, and Bryan Mound	Data unavailable.
Duval:	
Orchard	Leased from individuals. Fixed payment of $1.00 per ton for the first 6,000 tons produced each month and $2.00 per ton on all output beyond 6,000 tons per month. In addition, a fixed fee of $0.84 per ton is paid to the fee owners.

11. See "Sulphur Prices Hit New High," Chemical Week (7 October 1967), p. 32. Freeport Sulphur Company spent $14.8 million between 1965 and 1967 in searching for offshore Frasch deposits without success. Other firms reportedly spent $21 million on offshore exploration for Frasch sulphur during this period without finding a commercial deposit.

Table 19.   (Continued)

Company and mine	Description of royalties
Palangana and Boling Dome	Data unavailable.
Jefferson Lake: Lake Peigneur	Leased from the State of Louisiana. Royalty of $0.75 per ton paid to the state.
Clemens	Leased from individuals and subleased from Texas Gulf Sulphur. Amount of royalty unknown. Until 1949, payment to Texas Gulf was made in the form of sulphur. In 1949, royalty payments were substituted.
Long Point	Leased from Texas Gulf Sulphur. Texas Gulf receives 50 per cent of the output.
Starks	Leased from Texas Gulf Sulphur and Carter Oil Co. No data available on royalties.
Texas Gulf: Boling	Subleased from Gulf Production Co. (85.065 per cent), The Texas Co. (9.113 per cent), Sun Oil Co. (0.825 per cent), Vacuum Oil (1.972 per cent), and others (2.909 per cent). Royalty payments to Gulf are $1.00 per ton including payment of the underlying royalty; $3.00 per ton paid to The Texas Co. including the underlying royalty; and $1.00 per ton paid to the remaining interests plus the underlying royalty. Further, Texas Gulf agreed to share equally with the oil companies in the net profits after all development expenses were recouped. In 1934, Gulf Oil exchanged its interest in this property for 1.3 million shares of Texas Gulf stock or 38.83 per cent of the stock thereafter outstanding. On 3 November 1948, Texas Gulf repurchased 500,000 shares from Gulf Oil at $55 per share. Gulf Oil sold its remaining shares in 1956.
Long Point	Leased from Gulf Production Co. under the Boling Dome lease agreement.
Moss Bluff	Leased from Gulf Production Co. and Humble Oil and Refining Co. No data available on royalty payments.
Fannett, Gulf, and Spindletop	Data unavailable.

SOURCE: All data taken from Moody's *Manual of Industrial Investments*. Additional information taken from William Haynes, *Brimstone: The Stone That Burns* (Princeton, New Jersey: D. Van Nostrand Company, Inc., 1959).

NOTE: All sulphur mines developed prior to 1967 are included in the above listing except the original Sulphur Mine of Union, the Big Creek mine of Union, Standard Sulphur Company's Damon mine, and U.S. Sulphur Company's High Island mine. Data on these mines are unavailable.

royalty is paid when the Frasch producers sublease the sulphur rights from petroleum companies. This second royalty is sometimes paid to the petroleum company or companies in the form of a fixed payment per ton of sulphur mined, but more often is combined with a payment of a percentage of net profits from the operation.

Royalty payments are considered a cost of production, but when these payments amount to 50 per cent or more of the net profits from the operation, as shown in table 19 for several of the major deposits, the relationship between the lessor and lessee resembles that of a limited partnership. In such cases, royalties ought to be included in net income, and looked upon as part of the return on investment of capital in the firm. In the case of Boling Dome, Texas Gulf and Gulf Oil decided that an exchange of stock for sulphur rights would be preferable to a profit-sharing royalty agreement. In this case, royalties do not show up as a cost of operation (except for the royalty payments to the initial owners of the lease) but are included in net income and distributed to share-holders in the form of dividends. Of course, the stock given to the oil company in return for its deposit would represent an expansion of investment in the firm and the rate of return would reflect this expansion.

Royalty payments are not listed separately from other costs in the financial statements of Duval and Jefferson Lake. However, both Texas Gulf and Freeport list royalty payments as supplementary information to their profit and loss statements. Examination of royalty payments from 1934 through 1962, given in table 20, reveals two interesting facts. First, even though Texas Gulf was the larger producer during this period, Freeport paid two and one-half to three times more royalties. This is due to the fact that Texas Gulf's largest mine, Boling Dome, had been largely freed of royalty payments by the exchange of stock with Gulf Oil. A second point of interest is that royalty payments by both firms declined in the period 1957–62, reflecting the downward trend in output.

Cost-Price Relationships

Competitive performance would require that firms in an industry pass forward reductions in costs to buyers in the form of reductions in price; thus, prices would be closely related to the level of marginal costs. It is apparent from previous chapters that the Frasch sulphur industry does not have a competitive market structure. In the period prior to 1955, the Frasch sulphur industry was an oligopoly, selling a completely homogeneous product, and protected from entry by absolute cost barriers (stemming from possession of the lower-cost sulphur deposits). Thus, its prices at home and abroad reflected its market power and

Table 20. Royalty Payments by Texas Gulf Sulphur and Freeport Sulphur,
1934–62

thousand dollars

Year	Texas Gulf Sulphur*	Freeport Sulphur
1934	984	2,278
1935	963	2,751
1936	1,088	3,788
1937	1,149	4,435
1938	1,017	3,325
1939	656	3,083
1940	1,793	3,788
1941	2,173	4,469
1942	2,113	4,915
1943	2,094	5,332
1944	1,850	5,488
1945	1,850	6,116
1946	1,836	5,515
1947	2,216	5,001
1948	2,319	6,781
1949	2,211	8,266
1950	2,800	9,047
1951	3,385	10,493
1952	3,841	11,316
1953	4,013	13,858
1954	4,341	16,421
1955	4,945	17,329
1956	4,902	18,167
1957	4,501	15,296
1958	3,533	11,552
1959	3,798	11,468
1960	3,595	11,486
1961	3,403	11,134
1962	3,860	10,470

SOURCE: Moody's *Manual of Industrial Investments* for the applicable years.
* Includes both royalty payments on output and "advanced royalties" paid for renewal or extensions of rights.

yielded, as seen in this chapter, above normal profits for the industry. Prices were clearly not related to the level of marginal cost. Since 1955, the entry of new firms in response to expanding demand has significantly altered both the market structure and market behavior of Frasch sulphur firms. As might have been expected, the initial impact of entry was to reduce industry profits and cause major alterations in market behavior. Despite the changes in market structure and behavior that have occurred since 1955, the Frasch sulphur industry has remained an industry of few sellers producing a completely homogeneous commodity, but it has become increasingly difficult for Frasch sulphur firms to supply the entire market for elemental sulphur. As a result, higher-cost sources of supply have been brought into production. Sulphur prices now tend to reflect the higher costs of mining marginal deposits. The lower-cost sources of supply are thus earning pure economic rent. Therefore, in analyzing the relationship between costs and prices, it is necessary to examine the relationship as it existed prior to 1955, and as it exists today.

Cost-price relationships prior to 1955

Given the purely oligopolistic structure of the Frasch sulphur industry prior to 1955, the best explanation of industry pricing is that of tacit collusion on the domestic market and cartel pricing on the export market. Tacit collusion on the domestic market refers to the behavior of Texas Gulf and Freeport, acting as duopolists, in setting the domestic price that maximized industry profits. Tacit collusion, it should be stressed, means only that the dominant sellers in the market reached an implicit understanding regarding pricing and output behavior, not that they formally met or agreed to a common pricing policy. As the economic literature on oligopolistic pricing clearly reveals, this type of pricing behavior is the only way in which rational oligopolists can act if they are to avoid constant turmoil in their market.[12]

Chamberlin has shown that with only two firms in the market selling a homogeneous commodity and pricing independently, if both sellers recognize their mutual dependence and act accordingly, the industry price will stabilize where marginal cost is equal to marginal revenue, thus maximizing industry profits.[13] The stabilization of the domestic price of Frasch sulphur at $18.00 per ton f.o.b. the mine during 1925 and 1926 appears to correspond closely to the theoretical model of duopoly behavior.

Once established, such an oligopolistic price would be unlikely to change frequently. Where the degree of mutual dependence among the oligopolists is extreme, as is the case in an industry of very few sellers producing an undifferentiated product, prices are likely to be very sticky because each firm is uncertain of the reaction of its rivals to a change in price. This tendency toward price inflexibility would be reinforced in the case of Frasch sulphur by the extreme inelasticity of demand at prices below the profit maximizing level. (For example, it was shown in chapter 6 that, at a price of $14.00 per ton, Frasch sulphur would be used by all domestic acid manufacturers except those tied to

12. See Stigler, *The Theory of Price*, chap. 13; Martin Shubik, *Strategy and Market Structure* (New York: John Wiley and Sons, Inc., 1959); and H. H. Liebhafsky, *The Nature of Price Theory* (Rev. ed., Homewood, Illinois: The Dorsey Press, 1968), chap. 13.
13. Edward H. Chamberlin, *The Theory of Monopolistic Competition* (6th ed., Cambridge: Harvard University Press, 1950), pp. 46–51.

the use of another form of sulphur or those consuming by-product sulphur.) Small shifts in marginal costs would not cause the profit-maximizing price to depart significantly from the existing level. Finally, where entry is restricted by a high barrier—in this case, an absolute cost advantage—the initial determination of price is made with the purpose of limiting entry; upward adjustments in price would be resisted because they might induce unwanted entry into the industry. Thus, once the equilibrium price was established, strong forces would operate to prevent its change. Again, the behavior of domestic Frasch prices, which changed only once between 1926 and 1947, conforms closely to the behavior expected from firms in a purely oligopolistic industry.

Shifts in demand, however, if of significant magnitude, would cause the profit-maximizing price to change. However, the oligopolists would be likely to respond only if they became convinced that the shift in demand was to be permanent, and no firm would initiate a change in price until it was reasonably certain that its actions would be followed by its rivals. Thus, while prices would eventually respond to shifts in demand, the response would be slower than under a more competitive market structure. This analysis helps to explain the failure of the posted price of Frasch sulphur to fall in the face of the severe drop in demand caused by the depression of the thirties. Not until 1938 did Frasch producers reduce their posted price. It also helps to explain the rather sluggish response of Frasch sulphur prices to the rapid expansion in demand which occurred following World War II.

On the export market, the Webb-Pomerene Act made it legally possible for the two major Frasch producers to act jointly through the Sulphur Export Corporation in determining export prices. They responded by entering into a formal market-sharing agreement with the Sicilians. As discussed in chapter 5, the practice of price discrimination in export markets rested on the complete control over the supply of elemental sulphur. The cartel participants established export prices that reflected the costs of Sicilian sulphur, which were higher than those of Frasch sulphur. Thus, export prices were consistently higher than domestic prices, and sales in export markets were restricted to consumers requiring elemental sulphur. Even though the cartel agreements were abandoned by 1940, export prices continued to run above domestic prices throughout the existence of Sulexco.

Cost-price relationships after 1955

Higher-cost marginal Frasch mines, such as Bay Ste. Elaine, Damon, Starks, Chacahoula, and Nash, were brought into production after World War II when additional supplies were needed to meet the rapidly expanding demand. Prices for Frasch sulphur moved steadily upward from 1947 to 1956, interrupted only by the two-year period of governmental controls during the Korean crisis, and the low-cost mines of Texas Gulf and Freeport earned significantly higher economic rents. Beginning in 1955, U.S. Frasch sulphur met increasing competition both at home and abroad, first from Mexican Frasch sulphur and then from Canadian and French recovered sulphur. This increase in the supply of elemental sulphur brought about a reduction in sulphur prices, and resulted in the closing of many of the marginal mines.[14]

The period from 1955 to 1958 witnessed an apparent conflict in pricing objectives between Texas Gulf and Freeport. As described in chapter 6, the leadership efforts of Freeport in 1956 led to the abolishment of the export differential. This occurred during the six-year period 1952–58, in which the Sulphur Export Corporation did not exist. Freeport's actions, however, were followed by all other producers. An attempt by Texas Gulf in 1957 to reestablish the export differential met with opposition from Freeport and ended in a decrease in both domestic and export prices. Freeport emerged in a stronger position during this period because of its control of the interior, more protected domestic markets.

After 1958, the pricing structure of Frasch sulphur underwent a complete change as producers reestablished Sulexco, moved to liquid delivery, established regional storage terminals, and switched to a system of delivered pricing characterized by substantial concessions being made in terms of absorption of freight and handling charges in order to combat competition from other sources of sulphur. The experience of the Frasch sulphur industry during this period confirms the presence of competitive forces in the economy. The discovery of new sources of supply and the entry of new firms into the market were sufficient to upset the pattern of market conduct that had persisted for several decades. The developments in the Frasch sulphur industry bear out Fellner's conclusion that "The broadening of oligopolistic groups is likely to loosen the coordination of business policies, particularly with respect to market variables other than price."[15] While price reductions did occur following the entry of new sources of supply, the more important changes from the point of view of the long-run impact on the Frasch sulphur industry have been the significant alterations in both the price structure and distribution channels for Frasch sulphur.

---

14. The major reason given by Texas Gulf for abandoning its efforts to produce Frasch sulphur in Mexico in 1960 was the higher costs of the Mexican operation. Other mines abandoned prematurely during the period of declining sales realizations included Nash and Chacahoula. All of these mines were later reopened during the sulphur shortage of 1964–68, when increased prices made possible further production from these marginal deposits.

15. William Fellner, "Collusion and Its Limits Under Oligopoly," *American Economic Review* (May 1950), p. 60.

When rapid expansion in demand both in the United States and abroad beginning in the early sixties once more created a need for additional supplies of elemental sulphur, Frasch producers responded by first eliminating their special discounts and concessions, and then by instituting sizable increases in the posted price on both the domestic and export markets. However, prices continued to be established on the basis of delivered costs to regional markets. The pattern of the early fifties was repeated as higher prices once again brought forth additional elemental sulphur onto the market by the tapping of marginal sources of supply. The rents earned by the lower-cost Frasch producers again increased to sizable proportions.

It now appears that sulphur prices will be established at the level required to bring in the additional amounts of elemental sulphur needed to fill the gap between the demand for elemental sulphur in world markets and the available supply from low-cost sources such as Frasch deposits. The price level will be related not to the marginal costs of the Frasch producers, but to the marginal costs of mining lower quality sulphur deposits and utilizing non-Frasch sources of supply to the extent required to bring the market into balance. It is likely that Frasch producers will continue to earn a significant amount of economic rent. However, the regionalization of sulphur markets and the continued practice of delivered pricing provide more room for flexibility in sulphur pricing today than at any time in the past. The greater number of suppliers of elemental sulphur should help to ensure this flexibility.

Other Standards of Competitive Performance

In addition to the level of profits and the relationship between costs and prices, industry studies have employed a number of other indicators of competitive performance. These include: plant-output relations (the existence of plants of optimum scale and the absence of chronic excess capacity); selling expenditures (as a measure of the degree of product differentiation); and progressiveness (the existence over time of innovation and technological change). None of these standards is particularly suitable for analyzing the performance of the Frasch sulphur industry. The industry appears to have achieved reasonable long-run efficiency in scale and capacity utilization. Selling expenditures are not significant, as might be expected for an industry producing a homogeneous commodity that is purchased by industrial buyers. The industry appears to have been progressive, although it is difficult to determine a standard for comparison. Consideration of these factors would not appear to add much to our understanding of the Frasch sulphur industry's performance.

Chapter 8

The Industry and the Resource

The Frasch segment of the sulphur industry is engaged in exploitation of an exhaustible resource. In evaluating the industry's performance the question arises, how effective has the industry been in utilizing this resource? Has the mining been efficient? Has there been an adequate investment in the discovery and development of new resources? Have consumers been assured of an adequate supply? These are questions closely related to the market structure of the industry and the market behavior of Frasch sulphur producers.[1]

An equally important consideration is the outlook for sulphur as a resource, and, in particular, the supply-demand balance that may be expected to prevail in coming years. Can expected future demands on the sulphur resource be met without substantial increases in cost? What extensions of the resource base are likely to occur?

Effective Utilization of the Sulphur Resource

The Frasch sulphur firms seem to have been very efficient in exploiting the sulphur resource. In particular, the practice of exploiting each deposit as a unit under single management has avoided the waste so often associated with multiple ownership of a specific block of resources. On the other hand, until the past decade, exploration for new Frasch sulphur deposits was undertaken almost entirely by petroleum firms. Indeed, most of the Frasch deposits were discovered in the process of petroleum exploration. As discussed in chapter 7, exploration expenditures in the Frasch sulphur industry have been extremely low in relation to the average net income of the firms in the industry or to the amount of percentage depletion which the industry receives.

As the analysis of the industry's development has demonstrated, resource development takes place in response to needs that are transmitted through the market mechanism, either through changes in the

1. See Edward S. Mason, "The Political Economy of Resource Use," in *Perspectives on Conservation,* ed. Henry Jarrett (Baltimore: The Johns Hopkins Press for Resources for the Future, Inc., 1958), pp. 157–86.

supply conditions for a given resource or through changes in the demand for that resource. In general, increases in supply occur in response to prices that yield above-normal profits and lead to entry and the development of sources of supply that have not previously been competitive. Decreases in supply occur through the closing of marginal mines in response to prices which yield below-normal profits and lead to the exit of firms from the industry. Changes on the demand side occur in response to a change in relative factor costs, i.e., a change in the cost of a given resource input as compared with the cost of alternative resource inputs will encourage consumers to attempt to alter their consumption of the resource.

For resource development to be responsive to the market, it is essential that the market mechanism be free of artificial impediments. During the period from 1922 to 1955, sulphur prices remained high enough to yield an above-normal rate of return to Frasch sulphur producers. Changes in supply did not occur, however, primarily because ownership by the existing producers of the high-quality deposits made the cost of entry into the Frasch segment of the industry too high. The existence of an international cartel controlling the entire supply of elemental sulphur over much of this period also acted to restrict the market and to permit Frasch producers to engage in entry-preventing pricing. The price of Frasch sulphur was maintained below the price at which other forms of sulphur could profitably be produced, but above the level that would lead to normal profits for Frasch producers. At the same time, the output of Frasch sulphur during this period was generally in excess of the quantity demanded, with the major producers maintaining large stocks. Since Frasch sulphur was easy to obtain and its price was below that of alternative inputs, industrial consumers were not encouraged to develop sulphur-saving innovations or switch to substitute inputs.

The shortage of sulphur in the late forties and early fifties was the result of several factors. First, development of at least one new deposit was held up during World War II by a shortage of plant and equipment.[2] Second, the postwar expansion in demand could not have been foreseen. Finally, attempts by the government to maintain sulphur prices in the face of expanding demand discouraged resource development and led to a situation where available supplies of sulphur had to be rationed among consumers. Consumers then turned to other sources of sulphur (for example, gypsum in England) and developed sulphur-conserving processes. After Frasch sulphur prices were freed from artificial control, the market reacted quickly and prices increased. At this point, entry did occur with the development of other sources of supply, including

2. See Haynes, *Brimstone: The Stone That Burns*, pp. 162–64.

Frasch sulphur in Mexico, and recovered sulphur in Canada, France, and the United States.

A second shortage of sulphur occurred in the period from 1964 to 1967. Since the turn in the market began in mid-1963, it would appear that the industry had ample time to increase capacity. However, several factors must be considered before a judgment can be made. First, a part of the shortage was caused by the failure of Mexican Frasch exports to increase as anticipated. Second, Freeport was the only domestic Frasch producer that could increase output significantly in the short run. Third, the supply of recovered sulphur is inelastic in the short run owing to its by-product nature, and Canadian and French producers could not increase output immediately. Finally, because elemental sulphur had been in ample supply at low prices for several years, many consumers had switched from pyrites, and the surge in demand fell primarily on elemental sulphur. There was also government intervention in this period to prevent the domestic Frasch price from increasing. Without a price increase, there was less incentive to search for other sources of supply or to reduce sulphur consumption.

The lesson that emerges is that when an industry is based upon a depletable resource, public policy should have a broader focus than influencing the number of firms in the industry or the price level. At the same time, the oligopolistic structure of the industry, and its vulnerability to antitrust investigation, should not be used by the government to achieve price stability during periods in which the market is out of adjustment. Rather, emphasis should be placed on potential new sources of supply, new technologies, and new applications of existing techniques, and efforts should be made to ensure that resource development is not impeded by artificial barriers, such as selective price controls and export cartels.

The Outlook for Sulphur as a Resource

The sulphur market's performance in the next decade will depend upon the growth in demand, the extent to which the resource base can be expanded to meet this growth, and the resulting cost of sulphur to consumers. In the following sections, factors affecting each of these elements are discussed.

Growth in demand

The simplest method of projecting aggregate sulphur demand is to assume that the demand for sulphur will continue to grow at about the same rate as industrial production. A long-run annual growth rate of 4.4 per cent in industrial production, the rate actually experienced over the period from 1947 to 1957, would imply a rise in annual sulphur consumption in the United States from 8 million long tons in 1965 to 12.3 million long tons in 1975. However, such projections assume that the pattern of sulphur consumption will not change. To test the validity of that assumption, potential changes in the industrial consumption of sulphur are examined below, first for industries other than the fertilizer industry, and then for the fertilizer industry.

The pattern of industrial consumption of sulphur remained very stable for many years, the major components being chemicals, inorganic pigments, iron and steel pickling, production of man-made fibers, petroleum refining, and pulp and paper manufacture. In recent years, however, the use of substitutes for sulphur (or sulphuric acid) has reduced sulphur's share of some of these major industrial markets. One of the primary reasons for substitution has been the difficulty of disposing of spent sulphuric acid and the high cost of regeneration. As public tolerance of pollution has diminished, the search for less offensive alternatives to sulphuric acid has proceeded in many industries.

In iron and steel pickling, a component representing about 4 per cent of total domestic sulphur demand, the consumption of sulphuric acid increased 23 per cent between 1963 and 1965, while consumption of hydrochloric acid during the same period rose fivefold.[3] Of the 75 large picklers in the United States, 57 have decided to switch to hydrochloric acid.[4] The primary reason is that hydrochloric acid presents much less of a disposal problem. In addition, spent sulphuric acid is 8 to 10 per cent acid and too costly to regenerate, while hydrochloric liquor is less than 0.1 per cent acid, and with one of the newer processes about 80 per cent of the acid can be recovered from the pickling bath. Also, hydrochloric processes are said to give a better looking sheet than sulphuric acid processes and overpickling is less of a problem.

The rate at which hydrochloric acid will replace sulphuric acid in iron and steel pickling is difficult to estimate. The average life of a pickling line is fifteen years, which offers opportunities for replacement. However, even though pickling is much faster with hydrochloric acid than with sulphuric acid, some of the existing pickling plants are too slow to take full advantage of the speed of the hydrochloric process. Thus, while new plants will probably use hydrochloric processes, many of the existing plants will not be converted from sulphuric acid. The best estimate is that use of hydrochloric acid will increase to 700,000 tons by 1970, while the use of sulphuric acid will fall to about 400,000 tons. This would represent a decline of over 60 per cent in consumption by the iron and steel industry.[5]

3. *Chemical and Engineering News* (6 September 1965), p. 107.
4. "Hydrochloric Acid Dominates Steel Pickling," *Chemical and Engineering News* (3 March 1969), p. 38.
5. *Ibid.*

Another application in which sulphuric acid is facing stiff competition is in the manufacture of inorganic pigments. Traditionally, titanium dioxide pigments have been made by digesting ilmenite ore with sulphuric acid. Because of the varying content of iron and other impurities in ilmenite ores, the quantity of sulphuric acid used in digestion varies between 3 to 4 tons per ton of titanium dioxide pigment. Titanium slag may also be used to manufacture pigment and thereby acid requirements are reduced to 1.0 to 1.5 tons per ton of pigment. In 1965, about 560,000 tons of sulphur were used in the United States in the manufacture of titanium dioxide pigments, representing 7 per cent of the total domestic sulphur demand.

In the late fifties, DuPont introduced a process for manufacturing titanium dioxide pigments from rutile ore using chlorine. This process has gained wide acceptance in recent years because of several factors. Construction costs are about 60 to 70 per cent less than those for the sulphuric acid leaching process. Waste from the chlorine process is about one-eighth of that from sulphuric acid, and chlorine disposal problems are reduced because the chlorine process is continuous and the chlorine recycled. The chlorine process also has a small cost advantage, despite the fact that it can only be used with rutile ore which costs over twice as much as ilmenite or titanium slag.

Two additional components of sulphur demand which appear to be diminishing are petroleum refining and pulp and paper manufacture. Use of sulphuric acid in petroleum refining is decreasing due to the increasing application of solvent-extraction and hydrogenation. Alkylation, a major user in the oil refining industry, is not expanding, mainly due to the changeover in aviation fuel from high octane to kerosene type fuels and the growing application of hydrofluoric in place of sulphuric acid. In pulp and paper manufacture, the newer sulphate process, which uses sodium sulphate, is steadily replacing the sulphite process which uses sulphuric acid. The principal advantage offered by the sulphate process is the possibility of recovery and regeneration, and the smaller pollution problem.

Only in its application in the manufacture of chemicals and man-made fibers does sulphur appear to have a favorable outlook among major consumers. The use of sulphuric acid in the chemical industry is so widespread that curtailment in its use in the manufacture of any single product would probably have little effect on total consumption by the industry, and there is no indication that its use in many or most product lines will decline. In the synthetic fiber industry, however, a great deal will depend on the product mix in the industry. (Table 21 depicts the 1965 product mix in the industry and the resulting sulphur requirements.) In the next few years, it is likely that rayon's share of the total man-made fiber market will diminish in favor of such newer

Table 21.   The Use of Sulphur in Man-Made Fibers, 1965

Fiber	Fiber putput, 1965	Sulphur requirements
	million lbs.	*1,000 long tons*
Rayon	1,082	350
Acetate	445	22
Nylon 6 (Allied Chemical process)	712	132
Nylon 6/6 (DuPont process)	225	—
Polyester	361	3
Olefin	68	10
Acrylic	311	—
Spandex	8	—
Total		517

SOURCE: Interview with Delbert L. Rucker, Director of Information, The Sulphur Institute, Washington, D.C., 9 August 1966.

fibers as polyester and olefin which have much smaller sulphur requirements. In addition, at least two processes have been developed for the manufacture of rayon which use little or no sulphur. Nonetheless, barring a dramatic shift in product mix, the overall growth expected in the man-made fiber industry should mean that sulphur requirements will continue to expand.

In summary, the major characteristic of industrial consumption of sulphur other than by the fertilizer industry is that sulphur requirements will not increase as rapidly as demand for the products in which sulphur is used. Thus the share that sulphur holds in most of these industrial markets will continue to decline.

In recent years, the decline in certain industrial uses of sulphur has been more than offset by the increase in consumption of sulphur in the manufacture of fertilizers. For example, in 1963 fertilizer manufacture accounted for an estimated 40 per cent of total domestic sulphur consumption. It is estimated that about 77 per cent of the growth in demand for sulphur in the United States in the two ensuing years was for the fertilizer industry, with the result that by 1965, the fertilizer industry accounted for 45 per cent of total domestic sulphur consumption. It is likely, therefore, that the future growth in demand for sulphur will depend primarily upon the rate at which sulphur consumption expands in the manufacture of fertilizer.

The production of sulphur-containing fertilizers should continue to give strong impetus to growth of sulphur consumption over the next decade. Much of this growth will result from continued worldwide pressure on food production. Most of the world's agricultural lands including those in the United States can profitably utilize additional fertilization, and steep increases in fertilizer demand seem likely.

Sulphuric acid is used in the manufacture of both phosphatic and nitrogen fertilizers. (The amount of sulphur consumed in the manufacture of various types of fertilizer and the substantial growth in sulphur consumption by the fertilizer industry over the period from 1955 to 1965 are shown in table 22.) One of the most important factors behind the increased demand for sulphur for use in the manufacture of fertilizers has been the trend away from low-analysis fertilizers such as ammonium sulphate and normal superphosphate toward highly concentrated fertilizers such as triple superphosphate and ammonium phosphate. Since high-analysis fertilizers require up to 40 per cent more sulphur per unit of P_2O_5 in their manufacture, this trend has led to steadily increasing sulphur consumption by the fertilizer industry.

Table 22. Sulphur Used in Domestic Fertilizer Manufacture

Product	Analysis	Sulphur per ton	Sulphur consumption		
			1955	1962	1965
		lb.	(.... *million tons* ....)		
Ammonium sulphate	21% N	521	0.5	0.4	0.6
Normal superphosphate	20% P_2O_5	268	0.9	0.7	0.7
Triple superphosphate	45% P_2O_5	597	0.4	0.5	0.9
Ammonium phosphate	18% N 46% P_2O_5	880	0.1	0.5	1.0
Other	—	—	—	0.3	0.5
Total			1.9	2.4	3.7

SOURCE: L. B. Gittinger, Jr., Assistant Vice President, Freeport Sulphur Company, in a talk to the Chemical Marketing Research Association, 3 May 1966.

The sulphur used in the manufacture of the low-analysis products, ammonium sulphate and normal superphosphate, is an end component of these materials. In fact, there is more sulphur than nitrogen in a bag of ammonium sulphate and more sulphur than phosphorus in a bag of normal superphosphate. On the other hand, phosphoric acid, triple superphosphate, and ammonium phosphates contain little or no sulphur and the movement toward high-analysis fertilizers may soon create a need for direct addition of sulphur as a secondary plant nutrient. Soil sulphur deficiencies, which limit crop yields, have been noted in various parts of the world, including Australia, New Zealand, and Brazil, and at least sixteen states in the United States.[6]

6. See *Sulphur, The Essential Plant Food Element* (Washington, D.C.: The Sulphur Institute, 1962). Also see David W. Bixby, Samuel L. Tisdale, and Delbert R. Rucker, *Adding Plant Nutrient Sulphur to Fertilizer*, Technical

While sulphur consumption by the fertilizer industry increased substantially from 1955 to 1965, substitution of either fertilizers or processes not requiring sulphur inputs may dampen future increases in the demand for sulphur by this industry. An example of the first type of substitution is the use of non-sulphur nitrogen fertilizers such as urea, anhydrous ammonia, and ammonium nitrates in place of ammonium sulphate. Since each of these substitutes is of higher analysis than ammonium sulphate, this movement, which has been occurring on a large scale, is taking place regardless of the need or desire to lessen sulphur requirements.

Substitution of the second type is likely to come by the introduction of different methods of making phosphoric acid. Today, over one-third of the sulphur consumed in the United States enters into the production of wet-process phosphoric acid. The substitute most commonly suggested for sulphuric acid in the manufacture of fertilizers is nitric acid. "The nitric phosphates processes are the most attractive sulphur-saving methods . . . any further increase in the cost of sulphur, or a shortage of sulphur, could initiate a phenomenal swing toward nitric phosphates."[7] Annual production of nitric phosphates in European countries totals several million tons. In the United States, however, nitrophosphates are presently being manufactured in only six plants.[8] These plants use either a process that employs a mixture of nitric and sulphuric acids, the extract being ammoniated to produce diammonium phosphate, ammonium nitrate, and calcium sulphate, or a process using a mixture of nitric acid and phosphoric acid. Thus, in both types of plants, sulphur requirements are reduced, but not eliminated. The Norwegian Odda process, which is widely used in Europe, removes the calcium nitrate by refrigeration and crystallization and requires neither sulphuric nor phosphoric acid.

It is estimated that nitric phosphates currently have a small but significant cost advantage over ammonium phosphates and ammonium

---

Bulletin Number 10, October 1964, The Sulphur Institute. One of the primary sources of sulphur today is the gas released into the atmosphere by the burning of fuels such as coal. The sulphurous gases are captured by moisture in the air and are deposited in the soil by rainfall, fog, smog, and dew. It has been estimated that in the United States plants receive about 11 mil'ion tons of sulphur each year from the atmosphere. If strong pollution abatement measures were enacted nationally, about half of this sulphur could be recovered. However, soils would be deprived of sulphur required for plant growth, thus creating an enlarged market for direct application of sulphur as a plant nutrient.

7. Travis P. Hignett, Director of Chemical Development, Tennessee Valley Authority, in a paper presented at the National Plant Food Institute Annual Convention, White Sulphur Springs, West Virginia, 7–9 June 1965.

8. "Fertilizer Process Bypasses Sulphuric Acid Entirely," *Chemical and Engineering News* (20 November 1967), p. 19.

nitrate in large plants.[9] However, incomplete water solubility and inflexibility in N-P ratios continue to present a severe marketing problem. Further, the need for a large-scale plant limits the adaptability of most phosphoric acid operations to the use of nitric acid.[10]

The furnace method of producing phosphoric acid contains none of the product difficulties associated with the nitric acid substitution. However, costs appear to be prohibitive at present. It has been estimated that, even if electric power were free, the furnace method could not compete so long as sulphur costs less than $39.50 per long ton.[11] Of course, where a producer has surplus phosphorus from making other products, such as detergents, he can afford to use the excess to produce phosphoric acid.

At present, no producer is likely to build a plant to produce hydrochloric acid for use in fertilizer manufacture. However, if he is producing hydrochloric acid for some other type of manufacturing operation that requires it, he might well invest his surplus production in the manufacture of phosphatic fertilizers.

In summary, demand for fertilizers should continue to expand over the next decade, both within the United States and abroad. The most dynamic growth area for sulphur consumption by the fertilizer industry appears to be phosphate fertilizers, which already account for about 36 per cent of U.S. sulphur consumption and more than 30 per cent of the free world sulphur consumption. The trend toward use of high-analysis phosphate fertilizers should continue, while consumption of low-analysis products will have limited growth. Since high-analysis products require 40 to 50 per cent more sulphur in their manufacture than low-analysis products, the average annual growth of sulphur consumption by the fertilizer industry will be higher than the rate of increase in fertilizer output. Substitution possibilities, both product and process, appear unlikely to affect this outcome significantly over the next decade.

Supply outlook

The world will require new sources of supply to meet the continuing growth in demand. Fortunately, relatively abundant reserves of sulphur are available, though at higher production costs, and technology already

9. Hignett, *loc. cit.* This view was also supported by David Bixby, Chemical Engineer with The Sulphur Institute, in an interview in Washington, D.C., 3–4 August 1966.

10. A 600-ton-per-day nitrophosphate plant would require about 90 tons per day of nitrogen. An economically sized ammonia plant, however, would produce 820 tons per day of nitrogen.

11. Information obtained from David W. Bixby, Chemical Engineer with The Sulphur Institute. This assumes a plant making 54 per cent phosacid, and producing 200 short tons of P_2O_5 per day.

exists to tap these additional supplies. Further, as these new sources of supply become more important, competitive pressures should act to improve technology and lower the cost of recovery. Table 23 summarizes the world resource situation of sulphur. While it includes all sources of supply which have been identified, table 23 should not be taken to imply that all sources will be tapped. The outlook for each major source of sulphur is described below, and the figures given in the table are qualified as to origin and meaning.

The lowest-cost form of sulphur is Frasch sulphur obtained from the salt domes of the U.S. Gulf Coast and the anticlines of the Isthmus of Tehuantepec in Mexico. The U.S. Bureau of Mines estimates the reserves of known offshore and onshore U.S. salt dome deposits at 200 million long tons.[12] Intensive past exploration by the oil and gas industry suggests that additional sulphur-impregnated domes will not be discovered on the mainland at depths favorable for the recovery of sulphur. However, there are abandoned onshore Frasch mines that can be reworked, and plans to reopen five such mines were announced during the most recent shortage.

The offshore area of the Gulf Coast has not been thoroughly explored for sulphur, and the extent of potential offshore reserves may not be known for several years. It is likely that additional deposits of sulphur will be discovered in the offshore area, but they may prove too costly to mine to be classified with other salt dome mines as law-cost sources. Offshore exploration is estimated to be five times as costly as onshore exploration, and operating costs offshore probably run at least 1.5 to 2.5 times higher than those for an onshore mine of similar size and quality.

Estimated reserves in known Mexican anticline deposits on the Isthmus of Tehuantepec range from 50 million to 60 million tons.[13] However, Mexican sources indicate that the bulk of the Tehuantepec area has not been carefully surveyed by geologists. Further exploration on the Isthmus will probably result in the discovery of additional deposits of sulphur, but the irregular structure of the Isthmus formations makes it difficult to estimate the commercially recoverable quantities that might be added.

Sulphur ore also occurs in volcanic, hot spring, sedimentary, and replacement deposits. The sulphur content of these native ore deposits is

12. Paul M. Ambrose, "Sulphur and Pyrites," in *Mineral Facts and Problems, 1965 Edition* (U.S. Department of the Interior, Bureau of Mines, Bulletin 630), p. 909.

13. Marlin E. Sandlin, Chairman of the Board of Pan American Sulphur Company, has indicated the company estimates its reserves at 50 million tons. See *Chemical Week* (8 May 1965), p. 27. Gulf Sulphur is thought to have 7 to 10 million tons of reserves.

Table 23.   *Estimated World Sulphur Resources*

million long tons

Type and leading sources	Estimated resources
Elemental	
Frasch (salt dome)	250*
United States	200
Mexico	50*
Native	410*
United States	50*
Poland	100
Iraq	150*
Andes	75
Japan	35
Recovered	850*
United States:	
Gas	10*
Petroleum	45
Canada: Gas	350
France: Gas	35
Middle East:	
Gas	75
Petroleum	200
Rest of World:	
Gas	35
Petroleum	100
Oil shale, tar sands, and similar material	830*
United States	50
Canada	780
Nonelemental	
Metal sulphides	1,250*
United States	125
Spain: Pyrites	375
Other	750*
Anhydrite and gypsum	Enormous
Coal	15,000
United States	5,000
Sea water	Enormous

SOURCES: The information given represents a summary of available estimates of sulphur resources. It combines estimates of proven reserves with those of potential reserves. Where a range of estimates have been made, a representative figure has been arbitrarily selected. Those estimates thought to be conservative are noted. See Paul H. Ambrose, "Sulphur and Pyrites," *Mineral Facts and Problems, 1965 Edition* (U.S. Department of the Interior, Bureau of Mines, Bulletin 630), p. 909; "Sulphur Market Tight; Reserves Adequate," *The Oil and Gas Journal* (27 November 1967), pp. 71–74; and Arthur D. Little, Inc., *The Free World Sulphur Outlook* (April 1966).

NOTE: See the text section, "Supply outlook," for comments on the origin and meaning of the estimates in this table.

* Published estimates thought to be conservative.

similar to that of the Gulf Coast salt dome deposits—20 to 40 per cent—though occasional deposits contain up to 95 per cent sulphur. Most of these native ore deposits are located in remote areas, many of them are quite small, and production and transportation costs are higher than for either Frasch or recovered sulphur. In general, these sources are used to supply local markets. However, a large native sulphur deposit in the Mishraq area of northern Iraq is thought to be commercially feasible. Early in 1968, twelve companies were reported to be bidding for this deposit. The Iraq government was said to be demanding royalties of 80 per cent of net profits and requiring construction by 1972 of a million-ton-per-year plant and distribution facilities, estimated to cost about $40 million. The cost of sulphur from this deposit, delivered to the nearest port, Basra on the Persian Gulf, was estimated to be about $25 per ton.[14]

In the United States, small hot spring, volcanic, and sedimentary native sulphur deposits have been found in the Rocky Mountain and western states. Perhaps the most important domestic sulphur-bearing formations, outside the salt domes of the Gulf Coast, occur in Pecos County and Culbertson County, Texas. Two firms are attempting to mine the Pecos County deposit which is estimated to contain between 10 and 30 million tons of sulphur.[15] The Culbertson County deposit, being mined by the Duval Corporation, is estimated to have reserves of 50 million tons.[16]

The largest known deposits of volcanic ores occur in the 3,000 mile Andes Mountain chain in South America, where reserves have been estimated to contain as much as 100 million tons of sulphur.[17] The most important of the 100-odd deposits are found in Chile, where occurrences of high-grade (45 to 95 per cent) sulphur are estimated at 40 million tons. Similar deposits, generally having lower sulphur content, are known to exist in Peru, Argentina, Bolivia, Colombia, Ecuador, Mexico, Venezuela, Costa Rica, and the Galapagos Islands.

Japan has at least forty known deposits of volcanic sulphur with sulphur content estimated as ranging from 25 to 50 million tons.[18] Other Asiatic countries and many European countries also have volcanic

14. Thomas O'Hanlan, "The Great Sulphur Rush," *Fortune* (March 1968), p. 211.

15. F. F. Netzeband, Thomas R. Early, J. P. Ryan, and W. C. Miller, *Sulphur Resources and Production in Texas, Louisiana, Missouri, Oklahoma, Arkansas, Kansas, and Mississippi, and Markets for the Sulphur* (U.S. Department of the Interior, Bureau of Mines, Information Circular 8222, 1964), p. 11.

16. James C. Tanner, "Brimstone Supplies in 1968 Seen Equaling Consumption: Prices Start to Level Off," *The Wall Street Journal* (26 August 1968), p. 28.

17. Ambrose, *Mineral Facts and Problems, 1965 Edition*, p. 910.

18. W. T. Lundy, "Known and Potential Sulphur Resources of the World," *Industrial and Engineering Chemistry* (November 1950), p. 2200.

sulphur deposits, but most of them are too small and too remote to be considered as recoverable reserves under almost any conditions. The most important sedimentary deposits are found in the southeastern portion of Poland in the Tarnobrzeg basin. Reserves are estimated at 108 million long tons.[19] Open-pit mining is employed with flotation and filtration processing. Experiments are also underway with hot water and chemical solvent processes in an effort to reduce costs. The Solec-Grzybow deposit, a smaller but significant source located in the same area, is also being developed. Other areas having large sedimentary deposits include the U.S.S.R., Sicily, and mainland China.

The most important addition to the world's elemental sulphur reserves since the Korean War sulphur shortage has been the increased recovery of elemental sulphur from sour natural and refinery gases. Since the factors governing the availability of elemental sulphur from natural gas are quite different from those governing availability from petroleum, the two sources of recovered sulphur are considered separately below.

The sulphur content of U.S. natural gas averages about 0.05 per cent by weight. The sulphur content occurs as hydrogen sulphide in concentrations ranging from none to 70 per cent. In Canada, the concentration of hydrogen sulphide in gas presently being processed ranges from 1.0 to 38.1 per cent. Several of the planned recovery operations in the Middle East will be processing gas having a hydrogen sulphide content of 10 to 15 per cent. French plants at Lacq and nearby Mellion process gas which has a 5 to 15 per cent hydrogen sulphide concentration. Current technology enables the extraction of up to 95 per cent of the sulphur contained in the gas stream.

When it comes to estimating the reserves of sulphur that might be recovered from the gas fields, it is difficult to set a standard criterion for inclusion. Gas transported to fuel markets must be "sweetened" (the sulphur removed) if the hydrogen sulphide content is too high. If the gas contains more than 0.25 grain of sulphur per 100 cubic feet it is generally disqualified for sale to domestic consumers (0.25 grain is approximately 0.078 per cent sulphur by weight or 0.83 per cent hydrogen sulphide by weight). This suggests that natural gas fields with a sulphur content of over 0.078 per cent by weight should properly be considered as sulphur reserves *provided* a market exists for the gas, for if the gas is to be marketed the hydrogen sulphide must be removed. Natural gas fields having sulphur concentrations lower than 0.078 per cent by weight should not be included in reserve estimates unless sulphur prices increase substantially. Sour gas fields having concentrations

19. See "Poland Pushes Ambitious Chemical Plans," *Chemical and Engineering News* (5 October 1964), p. 65.

of sulphur above the maximum allowable but without access to markets must be treated individually.[20]

Reasonably accurate reserve estimates have been developed for sour gas fields in Canada, France, and the United States where sulphur is presently being recovered. For other areas of the world, only very rough approximations can be made. The U.S. Bureau of Mines estimates domestic sour gas sulphur reserves at 7 million tons from proven gas fields.[21] They might be five to ten times as high if they included gas not yet proven. The potential reserves of sulphur in the sour gas fields of Alberta have recently been estimated by the Canadian Petroleum Association at 350 million tons.[22] In making this estimate, the ratio of present proved sulphur reserves to proved reserves of natural gas with which they are associated was applied to the estimates of potential reserves of natural gas. Estimates of reserves at the two sour gas fields in France range from 25 to 50 million tons. Based on proven reserves in excess of 10 trillion cubic feet of gas, having an average concentration of hydrogen sulphide of 15 per cent, and allowing for 80 per cent recovery, an estimate of 35 million tons appears reasonable.

Middle East sour gas reserves are difficult to estimate. In 1965, Iran announced the discovery of a major sour gas field at Masjed Soleymān. Other sour gas fields exist in Iraq, Kuwait, and Saudi Arabia. Estimates of sulphur reserves in the Middle East range from 50 to 250 million tons, but a range of 50 to 100 million tons seems more realistic.[23] The distance of these fields from markets for natural gas means that only the development of fields with the highest hydrogen sulphide concentrations would be feasible. A recent study indicated that it might be more economical to import ammonia into the United States than to bring in LNG feedstock to make it here.[24] Such a development would have important implications for the development of sour gas fields in the Middle East. Gas from these fields might be processed to remove the natural gas liquids and sulphur for use in a local fertilizer complex. The gas could then be recycled into the formation, and the fertilizers exported. Thus, future development of sour gas fields in this area might not have to rely on a nearby market developing for the gas.

20. Arthur D. Little, Inc., *The Free World Sulphur Outlook* (April 1966), pp. 39–40.

21. Ambrose, *Mineral Facts and Problems, 1965 Edition*, p. 909.

22. U.S. Department of the Interior, Bureau of Mines, *Mineral Trade Notes* (July 1969), pp. 29–30.

23. Estimate made by H. W. Manley, president and managing director of Jefferson Lake Petrochemicals of Canada, Ltd., and cited in "Sulphur Market Tight; Reserves Adequate," *The Oil and Gas Journal* (27 November 1967), p. 73.

24. See "Ammonia Going One Way, Sulphur Another," *The Oil and Gas Journal* (16 October 1967), pp. 70–71.

Crude oils have widely varying sulphur content, ranging from 0.04 per cent (Pennsylvania) to 4.5 per cent (Mexico). In the United States the average sulphur content of crude petroleum is 0.6 per cent. Crudes containing higher amounts are usually blended with low-sulphur crude oils. In the past, because of the added difficulty of processing high-sulphur crude, many sources of high-sulphur crude were bypassed in favor of low-sulphur content crude. However, in recent years, it has become necessary to process crudes with an increasingly higher average sulphur content.

Refinery gases, but not heavy oil fractions, are generally desulphurized, and altogether only a small fraction of the contained sulphur is recovered from the crude. This leads to widely varying methods of estimating sulphur availability in crude oil. Thus, as with natural gas, estimates of recoverable sulphur contained in proven world petroleum reserves must be vague. It has been estimated conservatively that recoverable sulphur content in world proven oil reserves amounts to 330 million long tons.[25] Since proven oil reserves are a conservative measure, sulphur resources measured more broadly (including undiscovered pools, unproved portions of known fields, and secondary recovery) could be as much as 10 to 20 times the above estimate.

Large potential reserves of sulphur exist in the known deposits of other hydrocarbons such as bituminous rock, tar sands, and shale oil, but the range of estimates is very wide, depending on the estimator's judgment of the timing and feasibility of developing the hydrocarbon resource itself.

Operations began in 1967 to recover synthetic crude oil from the Athabasca tar sands in Alberta, which have an average sulphur content of 5 per cent by weight. Through February 1969, 13,165,761 barrels of raw bitumen had been processed, yielding 34,675 tons of sulphur. If future production is in the same ratio, the potential reserves of sulphur can be calculated at 780 million tons.[26] This estimate does not include the sulphur reserves of the Cold Lake oil sands in Northern Alberta, reported to contain 4 to 5 per cent sulphur, or those associated with certain structures in the Arctic Islands.[27] Based on estimated oil content in tar sands deposits of 2.6 billion barrels for the United States and 490 billion barrels for the world, and assuming a recovery factor

25. Arthur D. Little, Inc., *loc. cit.*, p. 41. Manley (cf. footnote 23) estimates proved reserves of sulphur in sour petroleum crude oil at 40 million tons and probable reserves at an additional 800 million tons.

26. U.S. Department of the Interior, Bureau of Mines, *Mineral Trade Notes* (July 1969), pp. 29–30. This is the estimate made by the Canadian Petroleum Association.

27. *Ibid.* Also, see "Where Are the Largest Native Sulphur Reserves?" *Oilweek* (7 April 1969), p. 23.

of 60 per cent, the sulphur content of bituminous rock and tar sands would be 10 million tons for the United States and 2 billion tons for the world.[28]

Shale oil, which to date has not been exploited in the United States, has an estimated sulphur content of 0.75 per cent. Assuming that estimated deposits could eventually yield 850 billion barrels for the United States and 1,297 billion barrels for the world, the sulphur contained in shale oil would be 846 million tons for the United States and 1,290 million tons for the world.[29] However, using a conservative approach toward the exploitability of shale, the Bureau of Mines places domestic sulphur reserves from this source at only 50 million tons.[30]

Metal sulphides, such as iron-bearing pyrites and various nonferrous ores that are smelted or refined for their copper, lead, and zinc content, are important sources of sulphur. These are widely distributed and can be found in all types of rocks. Practically all of the sulphur in nonferrous sulphides is released as sulphur dioxide in waste gas, and is normally recovered only when plants are located near sulphuric acid markets. Potential world reserves of sulphur in metal sulphides are estimated to be 1 to 1.5 billion tons.[31] The U.S. Bureau of Mines estimates that 100 to 150 million tons of sulphur content are available in domestic sulphides alone.[32] The principal producing countries for nonferrous sulphides are Japan, the United States, Canada, and the U.S.S.R. Spain has the largest reserves of pyrites, about 1 billion tons of ore. Other important sources are the U.S.S.R., China, Canada, Cyprus, France, West Germany, India, Italy, Japan, the Philippines, Norway, Portugal, Sweden, Turkey, Brazil, and North Korea. But given the very wide dispersal and the variety of conditions under which sulphur may or may not be recovered, these estimates actually list only rough orders of magnitude of eventually recoverable resources.

Sulphur may also be obtained from anhydrite and gypsum (calcium sulphate). Reserves of these sources are enormous and widely dispersed throughout the world. Gypsum is decomposed in the presence of silica, alumina, and carbon to produce sulphur dioxide and portland cement. The Elcor Chemical Company of Midland, Texas, is reported to have developed a process that permits the recovery of one ton of sulphur from about ten tons of gypsum, and the company has constructed a

28. Arthur D. Little, Inc., *loc. cit.*, p. 44.
29. *Ibid.*
30. Ambrose, in *Mineral Facts and Problems, 1965 Edition*, p. 909.
31. Lundy, *loc. cit.*, p. 2201, cites an estimate of world reserves of pyrites of 908 million tons, made in 1926 by the 14th International Geological Congress at Madrid, with the warning that subsequent discoveries may have caused reserves to increase by 50 per cent or more.
32. Ambrose, in *Mineral Facts and Problems, 1965 Edition*, p. 909.

350,000-ton-per-year recovery plant in Texas.[33] Gypsum and anhydrite are also reacted with ammonia and carbon to produce ammonium sulphate. In recent years, the United Kingdom has met its need for additional sulphur by using more anhydrite. For example, 700,000 tons of anhydrite can be substituted for 140,000 tons of sulphur for the manufacture of 400,000 short tons of sulphuric acid, with the residue from the decomposition of the anhydrite being used to manufacture cement.[34]

Coal is another substantial source of sulphur. At the present time, the sulphur content is generally discharged to the atmosphere as sulphur dioxide in stack gases, although some is recovered in Europe. According to the U.S. Bureau of Mines, the average sulphur content of U.S. coal is about 2.6 per cent. The sulphur content of recoverable U.S. coal reserves alone, taken at a conservative 220 billion tons, would come to 5 billion tons or more.[35] How much of the sulphur content will be recovered is another matter, depending not only on the amount of coal that will be burned, but on the fraction that will be considered as innocuous if left in the stack gas and on the economy of the recovery process.

The final source of sulphur is the sea. Sulphur is the fourth most abundant element in sea water following chlorine, sodium, and magnesium, but unless sulphur is obtained as a by-product in the production of rarer minerals from sea salts, there is no foreseeable economic justification for investment in the advanced technology required to extract sulphur from the sea.

About 4 billion tons of sulphur are potentially available from sources considered either to be economic or to approach economic exploitation levels, and some of these sources (Frasch, native, recovered, and pyrites) may prove to have substantially higher reserves. Coal, gypsum, anhydrite, and sea water provide an enormous backup potential which brings the sulphur content of known sources to over 20 billion tons. The question, then, is not whether or not the resource base can expand to meet increases in demand, but, rather, at what price additional supplies of sulphur will be forthcoming, and in what order.

Outlook implications

This brief survey of the demand and supply outlook for sulphur, coupled with the earlier analysis of the industry's development, makes

33. O'Hanlan, *loc. cit.*, p. 212. Also, see "Trouble Ahead for Non-Frasch Sulphur?" *Chemical Week* (28 December 1968), p. 8, where Elcor officials are quoted as saying that there is enough gypsum on the company's 28,000 acre tract to last 100 years at the present rate of mining.

34. See "Acid from Indigenous Anhydrite," *Chemical Trade Journal and Chemical Engineer* (London: 20 May 1965), p. 618.

35. Ambrose, in *Mineral Facts and Problems, 1965 Edition*, p. 909.

possible some general conclusions regarding the future course of the industry. Demand for sulphur should continue to expand over the next decade at a rate equal to, or perhaps slightly above, the rate of growth in industrial production, depending on the degree to which losses in some of its industrial markets are offset by increases in the demand for sulphur for use in fertilizers. On the supply side, there are ample reserves of low-cost sulphur (including both domestic and foreign Frasch and recovered sulphur) to ensure that forthcoming demands will be met notwithstanding temporary dislocations. However, *with present recovery techniques,* supplies may become available at slightly higher prices than those that prevailed from 1926 to 1964. Moreover, the geographical dispersion of the primary sources of supply will tend to create regional markets for sulphur, differentiated by differences in freight costs.

At the same time, prices for sulphur should become increasingly the result of competition. The domestic Frasch industry can no longer supply the world's elemental sulphur needs. The new sources of supply have created a competitive environment for this industry. Domestic and foreign Frasch producers will continue to earn an economic rent as a reward for possession of the lowest-cost sources of supply, but the amount of that rent should diminish over time as these low-cost sources are depleted.

To anticipate both a more competitive price structure for sulphur and higher prices for sulphur consumers as well may appear paradoxical. However, if changes in technology lead to lower recovery costs, the more competitive environment will help to ensure that prices to consumers will also be lower.

Chapter 9

Summary and Conclusions

Analysis of the market structure, behavior, and performance of the Frasch sulphur industry was undertaken with the view in mind of investigating the use of market power and evaluating the implications of such use for the public welfare. The purpose of this chapter is to summarize the important aspects of the domestic sulphur industry's market structure, behavior, and performance and to set forth the author's conclusions regarding the workability of competition in the industry.

Characteristics of the Industry—Summary

Market structure

1. The U.S. Frasch sulphur industry came into existence at the turn of the century when a unique technological innovation gave producers using the Frasch process a significant cost advantage over the Sicilian native sulphur industry, which had been supplying most of the world's elemental sulphur. (Chapter 2, pp. 11–14.)

2. The U.S. Frasch sulphur industry has been one of few sellers. From 1896 to 1912, Union Sulphur Company enjoyed a complete monopoly on Frasch sulphur production by virtue of the patent granted to its founder, Dr. Herman Frasch. In 1912, Freeport Sulphur Company entered the industry, followed in 1919 by Texas Gulf Sulphur Company. In 1925, depletion of its mine forced Union to cease production. In 1928, Duval Texas Sulphur Company entered the industry, followed four years later by Jefferson Lake Sulphur Company. From 1932 to 1968, these four companies controlled the entire Frasch sulphur output of the United States, with Texas Gulf and Freeport accounting for nearly 90 per cent. The industry is not only concentrated, but production is centralized in a few mines located within a small region of the Texas and Louisana Gulf Coast. (Chapter 2, pp. 14–20.)

3. Frasch sulphur currently accounts for about 75 per cent of the

145

total output of sulphur in the United States, and about 85 per cent of domestic production of elemental sulphur. Most of the other U.S. elemental sulphur is recovered from sour natural or refinery gases. The primary form of nonelemental sulphur is pyrites. While all forms of elemental sulphur are perfect substitutes for one another, the degree of substitutability between elemental and nonelemental sulphur is limited. (Chapter 1, pp. 1–2.)

4. Foreign sources currently account for about 47 per cent of the total output of elemental sulphur in non-Communist countries, Mexico, Canada, and France having become important suppliers in the 1950s. (Chapter 2, pp. 20–33.)

5. Sulphur is used in an extremely large number of chemical and manufacturing processes, predominantly in the form of sulphuric acid, the production of which currently absorbs 86 per cent of sulphur output. There is no formal commodity exchange for sulphur. In general, it is sold on a contract basis directly to industrial consumers. Two important changes in sulphur marketing took place in 1959: a switch from solid to liquid delivery of Frasch sulphur, accompanied by an expansion in the number of services provided for customers; and a change from prices quoted f.o.b. mines or Gulf port to pricing on a delivered basis at regional terminals. Over most of its history, the Frasch sulphur industry has handled the bulk of its exports through a Webb-Pomerene Export Association, the Sulphur Export Corporation (Sulexco). (Chapter 4, pp. 50–52 and 55–64.)

6. The demand for Frasch sulphur is derived from the demand for the products in which it is used. Demand is inelastic in the short run, but becomes more elastic in the long run since there are few unique uses for sulphur. Demand for Frasch sulphur is subject to only moderate seasonal fluctuations. In the past there has been a strong correlation between demand for sulphur and the index of industrial production, reflecting the pervasive use of sulphur throughout the industrial sector of the economy. Thus, demand is responsive to cyclical fluctuations in business conditions, and expands when industrial production rises. (Chapter 4, pp. 52–55 and 64–66.)

7. Investment costs per unit of capacity decline significantly with increases in the scale of plant. Engineering cost analysis reveals that about 75 per cent of total production costs are variable in nature. The crucial determinant of cost is the water ratio, i.e., the number of gallons of hot water required to produce a ton of sulphur. Water ratios vary with the quality of the deposit being worked and range from as low as 1,000 gallons to as high as 12,000 gallons of hot water per ton of sulphur recovered. Plants having identical water ratios experience decreasing average unit costs with increases in plant scale over the range of plant sizes currently found in the industry. By the same token, the

average unit costs of plants of identical size were found to vary directly with the water ratio. However, the water ratio is more important than plant size in determining the level of average unit costs. Thus, Frasch sulphur unit production costs reflect the existence of substantial internal economies of scale, resulting from savings realized in the use of fixed factors as plant size increases. When the industry is operating at full capacity, it is not possible to increase output in the short run without increasing average costs per ton. (Chapter 3, pp. 34–42 and 47–49.)

8. Entry into the Frasch sulphur industry is very difficult, the principal barrier being one of absolute cost resulting from control by the established producers of the better deposits. Neither product differentiation nor economies of scale seemed to constitute a major barrier to entry. (Chapter 3, pp. 42–47.)

Market behavior

9. Shipments of Frasch sulphur are subject to some seasonal variation. However, seasonal and irregular fluctuations in monthly consumption were found to be unimportant in earlier years because producers maintained extraordinarily large inventories. Production schedules were found to lag behind demand changes by as much as six months to a year. Inventories began to decline after producers had shifted in 1959 to liquid delivery of sulphur, and from 1962 to 1968 shipments exceeded production. (Chapter 5, pp. 69–74.)

10. Frasch sulphur production has grown steadily. While movements in production roughly coincide with cyclical movements in industrial activity, the correspondence is far from exact, particularly in the past decade. (Chapter 5, pp. 74–75.)

11. Frasch sulphur prices were found to belong to the class of prices commonly termed "administered"—they did not move to equate short-run demand and supply. Posted prices changed only once between 1926 and 1947, and only seven times between 1926 and 1964. In recent years, sulphur prices have risen to the highest levels in the industry's history in the face of a worldwide shortage of sulphur. There is no general trend in Frasch sulphur prices over the history of the industry, though definable shorter-term elements can be isolated. Net realized prices for exports were consistently higher than those for domestic shipments. In general, however, the direction and timing of movements in net realized prices for both export and domestic shipments coincided with movements in the posted price. The only period in which major departures were noted was between 1957 and 1963 when posted prices remained stable, but net realizations on both export and domestic shipments continued to decline. (Chapter 5, pp. 75–78, Chapter 6, fig. 13, p. 98.)

12. Price discrimination between the domestic and export markets was found to have existed over the industry's history with the exception of the period from 1960 to 1964. Higher prices on the export market prior to 1940 were found to have been closely related to the cartel agreements reached by the Sulphur Export Corporation and major foreign producers of elemental sulphur. (Chapter 5, pp. 78–82.)

Performance

13. Analysis of the rate of return earned by the industry on average invested capital revealed an abnormally high rate of return averaging 23.60 per cent from 1919 to 1953. While data on profits and invested capital pertaining to Frasch operations alone are only partially complete for the period since 1953, evidence indicates that rates of return remained above normal until about 1955 and then fell consistently until 1963. Investigation also disclosed that the average rates of return for the four firms in the industry had ranged from a high of 26.35 per cent for Texas Gulf to a low of 13.68 per cent for Freeport. (Chapter 7, pp. 111–17.)

14. The high rates of return in the sulphur industry were shown to have prevailed despite the fact that large royalty payments were made to the owners of the mineral rights to salt dome properties, mainly large oil companies. The 23 per cent depletion allowance on sulphur contributed to the high earnings of firms in the industry. But because sulphur deposits are almost always found in the search for oil, the effectiveness of the depletion allowance as an incentive to search for additional sulphur reserves has been modest, as evidenced by the fact that exploration expenditures in the past have been low in relation to the net income of firms in the industry and the amount of depletion which the industry receives. (Chapter 7, pp. 117–21.)

15. Frasch sulphur prices have not in general been closely related to the level of marginal cost. Prior to 1955, the Frasch sulphur industry was a pure open oligopoly (few firms, homogeneous product, high barriers to entry), and domestic prices were established to maximize industry profits. Export prices were determined by the Sulphur Export Corporation and, from 1926 to 1940, Sulexco set export prices through a cartel agreement with the Sicilian producers. Export prices exceeded domestic prices in part reflecting the higher marginal costs of the Sicilians. After 1955, the entry into the industry of two Mexican Frasch producers and the development of competition from recovered sulphur production in Canada and France significantly altered the market structure and behavior of Frasch firms. The expansion in demand in recent years resulted in marginal costs becoming more important in the determination of Frasch prices. However, it is the marginal costs of pro-

ducing sulphur from lower quality deposits and non-Frasch sources that determine Frasch prices. The major Frasch producers are able to earn significant amounts of economic rent. (Chapter 7, pp. 121–26.)

16. The industry was found to have utilized the sulphur resource effectively. In particular, the practice of unit management of deposits has fostered efficient mining practices. (Chapter 8, pp. 127–29.)

17. The outlook for sulphur as a resource is for continuing growth in demand, with increased consumption of sulphur in fertilizers more than offsetting a loss of market share in some industrial markets. The supply of sulphur should be adequate to meet foreseeable demands, but increased supplies will become available at prices slightly above those prevailing over the period from 1926 to 1964 unless there are improvements over current recovery techniques. At the same time, increased competition among sources should exert downward pressure on sulphur prices. (Chapter 8, pp. 129–44.)

Is the Frasch Sulphur Industry Workably Competitive?

As pointed out in the introductory chapter, the key elements in any evaluation of the workability of competition in an industry are the degree to which actual performance has departed from desired performance and the possibility of remedial public action. Over most of its life the Frasch sulphur industry has not performed in a competitive manner. The record of stable prices and soaring profits in the face of a constant expansion in investment in the industry clearly indicates the dominance of monopolistic elements. These elements have been dormant in recent years and are unlikely to reappear now that there is competition from other sources of sulphur.

The scarcity of low-cost Frasch sulphur deposits has limited the industry to a few firms, and the differential quality of the workable deposits gives power over the market to at best one or two firms. The strongest factor working to make the Frasch industry perform competitively is the existence of other sources of sulphur. Much of the competitive pressure from other sources of supply has come from foreign producers.

In recent years, the development of the Frasch deposits of Mexico and the growth in recovery of sulphur from sour natural gas in Western Canada and France, have challenged the market power held by domestic Frasch producers. Because sulphur could be produced from these sources at a cost competitive with U.S. Frasch sulphur produced on the Gulf Coast, the foreign suppliers were able to bring pressure to bear on both the domestic and export markets. As shown in this study, this pressure resulted in lower prices, increased customer services, and reduced profits for domestic Frasch producers. Although prices and profits recovered before long and have recently climbed to historic highs, it is unlikely that the Frasch industry can ever return to its pre-1957 structure. Other sources of supply, both domestic and foreign, should become increasingly important over the next decade.

It is perhaps unlikely that the Frasch sulphur industry can be made truly competitive. In the long run, the fate of competition in the sulphur industry will rest on the expansion of the resource base. The discovery of new sources of supply, innovations reducing the cost of existing sources of supply, and the development of substitutes for use in some of the processes that currently require sulphur offer the best possibilities for protecting the interests of the public. The task of public policy should be to make certain that the market is sufficiently free of artificial restraints to enable these forces to exert their full influence.

Appendixes

Statistics on Production, Trade, Consumption, and Prices

Table A-1. U.S. Sulphur Output from All Sources, 1880 to 1967

thousand long tons

Year	Native sulphur	By-product sulphur				Pyrites		Total
		Gas & coal	Cu, Zn mining	Hydrogen sulphide	By-product total	Ore	Sulphur content	
	(1)	(2)	(3)	(4)	(5)	(6)	(7)	(8)
1880	0.54					2.0	0.9	1.8
1881	.54					10.0	4.5	6.4
1882	.54					12.0	5.4	7.6
1883	.89					25.0	11.0	15.0
1884	.45					35.0	16.0	20.0
1885	.64					49.0	22.0	30.0
1886	2.23					55.0	25.0	35.0
1887	2.68					52.0	23.0	34.0
1888	0					54.3	24.0	31.0
1889	.40					93.7	42.0	54.0
1890	0					99.9	45.0	58.0
1891	1.07					106.0	48.0	63.0
1892	2.40					110.0	50.0	67.0
1893	1.07					75.9	34.0	45.0
1894	.45					106.0	48.0	62.0
1895	1.61					99.5	45.0	61.0
1896	4.70					116.0	52.0	74.0
1897	2.03					143.0	64.0	85.0
1898	1.07					193.0	87.0	114.0
1899	4.31					175.0	79.0	107.0
1900	3.15					205.0	92.0	122.0
1901	6.87					235.0	106.0	145.0
1902	7.44					200.0	90.0	125.0
1903	7.38					226.0	102.0	140.0
1904	85.00					207.0	93.0	232.0
1905	220.00					253.0	114.0	425.0
1906	295.00					261.0	117.0	530.0
1907	189.00					247.0	111.0	390.0
1908	364.00					223.0	100.0	590.0
1909	274.00					247.0	111.0	490.0
1910	247					242	109	460
1911	205		99		100	302	140	440
1912	788		139		140	351	160	1,090
1913	491		143		140	341	150	780
1914	418		173		170	337	150	740
1915	521		205		200	394	180	900
1916	650		264		260	439	200	1,110
1917	1,134		330		330	483	220	1,680
1918	1,354		n.a.		255	464	210	1,820
1919	1,191		184		180	421	170	1,540

Table A-1. (Continued)

		By-product sulphur				Pyrites		
Year	Native sulphur	Gas & coal	Cu, Zn mining	Hydrogen sulphide	By-product total	Ore	Sulphur content	Total
	(1)	(2)	(3)	(4)	(5)	(6)	(7)	(8)
1920	1,255		279		280	311	120	1,660
1921	1,880		140		140	157	70	2,090
1922	1,831		n.a.		160	173	80	2,070
1923	2,036		178		180	191	80	2,300
1924	1,221		198		200	168	70	1,490
1925	1,409		228		230	194	80	1,720
1926	1,890		236		240	227	70	2,220
1927	2,112	2.5	236		240	303	120	2,470
1928	1,982	2.5	264		270	313	113	2,365
1929	2,362	2.5	286		290	334	121	2,773
1930	2,559	2.5	269		272	343	124	2,955
1931	2,129	2.5	196		198	331	121	2,448
1932	890	2.5	136		138	190	66	1,094
1933	1,406	2.5	139		142	284	108	1,656
1934	1,422	1.5	131		132	432	168	1,722
1935	1,633	1.5	137		139	514	203	1,975
1936	2,016	1.5	166		167	547	217	2,400
1937	2,742	1.5	189		191	584	232	3,165
1938	2,393	4.0	156		160	556	219	2,772
1939	2,091	4.0	177	13	194	519	220	2,505
1940	2,732	4	191	16	211	627	262	3,205
1941	3,139	5	208	21	234	645	270	3,643
1942	3,461	5	218	19	242	720	307	4,010
1943	2,540	5	279	18	302	802	337	3,179
1944	3,219	19	263	22	304	789	333	3,856
1945	3,754	25	246	19	290	723	296	4,340
1946	3,862	35	209	18	262	813	337	4,461
1947	4,443	43	212	21	276	941	392	5,111
1948	4,870	44	187	26	257	929	388	5,515
1949	4,747	57	167	38	262	888	378	5,387
1950	5,193	142	216	42	400	931	393	5,986
1951	5,280	184	241	60	485	1,018	433	6,198
1952	5,295	252	253	67	572	994	418	6,285
1953	5,194	342	253	80	675	923	380	6,249
1954	5,579	359	259	73	691	909	405	6,675
1955	5,800	399	325	94	818	1,007	410	7,028
1956	6,484	465	348	89	902	1,070	432	7,818
1957	5,579	511	391	88	990	1,067	436	7,005
1958	4,646	640	360	92	1,092	974	403	6,141
1959	4,640	686	317	88	1,091	1,057	437	6,168
1960	5,037	767	345	95	1,207	1,016	416	6,660
1961	5,477	858	332	106	1,296	987	399	7,172
1962	5,025	900	355	98	1,353	916	379	6,757
1963	4,882	947	356	116	1,419	825	344	6,645
1964	5,228	1,021	366	123	1,510	847	354	7,092
1965	6,116	1,215	388	139	1,742	875	354	8,212
1966	7,001	1,240	424	134	1,798	872	356	10,953
1967	7,014	1,268	364	134	1,766	861	355	10,901

SOURCES: 1880–1957 data from Neal Potter and Francis T. Christy, Jr., *Trends in Natural Resource Commodities* (Baltimore: The Johns Hopkins Press for Resources for the Future, Inc., 1962), pp. 417–18. 1958–67 data from U.S. Department of the Interior, Bureau of Mines, *Minerals Yearbook*, vol. 1.

n.a.—Not available.

Table A-2.   *U.S. Imports, Exports, and Consumption of Sulphur, 1880 to 1967*

thousand long tons

Year	Native sulphur imports	Native sulphur exports	Native sulphur shipments	Apparent consumption of native sulphur	Sulphur content of pyrites output	Sulphur content of pyrites imports	By-product sulphur output	Apparent consumption of all forms of sulphur
	(1)	(2)	(3)	(4)	(5)	(6)	(7)	(8)
1880	88		1	89	1		10	110
1881	105		1	106	5		10	130
1882	98		1	99	5		10	130
1883	95		1	96	11		20	140
1884	105		*	105	16	8	20	150
1885	97		1	98	22	3	20	140
1886	118		2	120	25	1	20	170
1887	97		3	100	23	8	20	150
1888	98		0	98	24	n.a.	30	150
1889	136		*	136	42	n.a.	30	210
1890	163		0	163	45	n.a.	30	240
1891	117		1	118	48	45	40	250
1892	101		2	103	50	69	40	260
1893	106		1	107	34	88	40	270
1894	125		*	125	48	75	40	290
1895	122		2	124	45	86	40	290
1896	139		5	144	52	90	50	340
1897	137		2	139	64	117	50	370
1898	151		1	152	87	114	60	410
1899	140		4	144	79	121	60	400
1900	167		3	170	92	145	70	480
1901	175		7	182	106	182	70	540
1902	171		7	178	90	198	80	550
1903	191		7	198	102	189	80	570
1904	130	3	85	212	93	190	90	580
1905	84	12	220	292	114	230	100	740
1906	74	14	295	355	117	270	100	840
1907	23	36	272	259	111	280	100	750
1908	21	28	206	199	100	300	100	700
1909	31	37	258	252	111	310	120	790
1910	31	31	251	251	109	360	120	840
1911	29	28	254	253	140	450	100	940
1912	30	58	305	277	160	440	140	1,020
1913	23	89	319	253	150	380	140	920
1914	26	98	342	270	150	460	170	1,050
1915	25	37	294	282	180	430	200	1,090
1916	21	129	767	659	200	560	260	1,680
1917	1	153	1,120	968	220	435	330	1,950
1918	*	131	1,267	1,136	210	220	255	1,820
1919	*	225	678	453	170	175	180	980
1920	*	477	1,518	1,041	120	150	280	1,590
1921	*	286	954	668	70	98	140	980
1922	*	486	1,344	858	80	125	160	1,220
1923	*	473	1,619	1,146	80	120	180	1,530
1924	1	482	1,537	1,056	70	110	200	1,440
1925	*	629	1,858	1,226	80	120	230	1,660
1926	*	577	2,073	1,490	70	160	240	1,960
1927	3	789	2,072	1,272	120	110	240	1,740
1928	5	685	2,083	1,383	113	210	270	1,980
1929	1	855	2,437	1,566	121	230	290	2,210
1930	*	609		1,381	124	160	272	1,940
1931	0	420		957	121	160	198	1,440
1932	0	360		749	66	110	138	1,060
1933	5	531		1,111	108	170	142	1,530
1934	6	517		1,102	168	160	132	1,560
1935	2	413		1,223	203	180	139	1,740
1936	1	567		1,403	217	190	167	1,980
1937	1	689		1,778	232	235	191	2,440
1938	3	592		1,040	219	150	160	1,570
1939	14	653		1,595	220	215	194	2,220

Table A-2. (Continued)

Year	Native sulphur imports	Native sulphur exports	Native sulphur shipments	Apparent consumption of native sulphur	Sulphur content of pyrites output	Sulphur content of pyrites imports	By-product sulphur output	Apparent consumption of all forms of sulphur
	(1)	(2)	(3)	(4)	(5)	(6)	(7)	(8)
1940	28	766		1,820	262	180	211	2,470
1941	29	761		2,344	270	165	234	3,010
1942	26	586		2,472	307	135	242	3,160
1943	17	682		2,528	337	115	302	3,280
1944	*	675		2,905	333	80	304	3,620
1945	*	943		2,907	296	85	290	3,580
1946	*	1,246		2,848	337	88	262	3,530
1947	*	1,350		3,490	392	61	276	4,220
1948	*	1,296		3,720	388	52	257	4,420
1949	*	1,461		3,410	378	58	262	4,110
1950	*	1,479		4,158	393	100	400	5,050
1951	2	1,312		3,786	433	106	485	4,810
1952	5	1,338		3,728	418	142	572	4,860
1953	1	1,271		3,932	380	91	675	5,080
1954	1	1,675		3,700	405	134	691	4,930
1955	35	1,636		4,246	410	171	818	5,640
1956	212	1,675		4,268	432	175	902	5,780
1957	499	1,593		3,997	436	169	990	5,590
1958	591	1,602		3,652	403	164	1,092	5,311
1959	642	1,636		4,232	437	134	1,091	5,894
1960	741	1,787		3,950	416	146	1,207	5,719
1961	832	1,596		3,908	399	135	1,296	5,738
1962	1,040	1,554		4,065	379	145	1,353	5,942
1963	1,351	1,612		4,302	344	93	1,419	6,158
1964	1,462	1,928		4,738	354	120	1,510	6,722
1965	1,486	2,624		5,134	354	160	1,742	7,390
1966	1,514	2,326		6,113	356	160	1,798	8,427
1967	1,474	2,043		6,260	355	165	1,766	8,546

SOURCES: 1880–1957 data from Neal Potter and Francis T. Christy, Jr., *Trends in Natural Resource Commodities* (Baltimore: The Johns Hopkins Press for Resources for the Future, Inc., 1962), pp. 490–91. 1958–67 data from U.S. Department of the Interior, Bureau of Mines, *Minerals Yearbook*, vol. 1.

NOTE: Column 4 is derived from columns 1, 2, and 3 through 1924, but represents a Bureau of Mines independently calculated series beginning in 1925. Column 8 represents a rounded addition of columns 5, 6, and 7, adjusted in 1880–83 to allow for lack of data.

* Less than 500 long tons.

n.a. Not available.

Table A-3.   Average Revenue and Posted Price per Ton, U.S. Shipments of Sulphur, Domestic and Export, 1900 to 1967

	Total value			Average value			Average posted price (7)
Year	All shipments (1)	Export shipments (2)	Domestic shipments (3)	All shipments (4)	Export shipments (5)	Domestic shipments (6)	
	(........ thousand dollars ....)			(.............. dollars/long ton ....)			
1900							21.15
1901							22.00
1902							23.45
1903							22.34
1904	1,256			20.90			21.79
1905	3,305			20.40			21.28
1906	3,702	298	3,404	17.30	20.67	19.95	22.16
1907	4,771	735	4,036	17.50	20.45	17.11	21.50
1908	3,727	562	3,165	18.10	20.13	17.72	21.81
1909	4,782	737	4,045	18.50	19.84	18.30	22.00
1910	4,522	533	3,969	18.00	17.99	18.03	22.00
1911	4,573	545	4,028	18.00	19.41	17.85	22.00
1912	5,289	1,076	4,213	17.30	18.64	17.01	22.00
1913	5,617	1,600	4,017	17.60	17.93	17.46	22.00
1914	6,214	1,807	4,407	18.20	18.41	18.07	22.00
1915	4,959	725	4,234	16.90	19.42	16.50	22.00
1916	12,246	2,506	9,740	16.00	19.46	15.26	31.36
1917	23,987	3,501	20,486	21.40	22.92	21.17	43.33
1918	27,868	3,627	24,241	22.00	27.66	21.35	28.62
1919	10,252	6,326	3,926	15.10	28.15	21.18	28.00
1920	30,000	8,994	21,006	20.00	18.84	20.19	23.85
1921	17,000	4,525	12,475	17.80	15.83	18.66	25.80
1922	22,000	7,096	14,904	16.40	14.61	17.37	14.08
1923	26,000	7,105	18,895	16.10	15.04	16.48	14.00
1924	25,000	7,793	17,207	16.30	16.16	16.31	14.02
1925	29,000	11,000	18,000	15.60	17.48	14.65	14.69
1926	37,300	10,918	26,382	18.00	18.92	17.64	18.22
1927	38,300	16,254	22,046	18.48	20.59	17.05	18.00
1928	37,500	14,345	23,155	18.00	20.94	16.56	18.00
1929	43,800	17,629	26,171	17.97	20.61	16.54	18.00
1930	35,800	12,416	23,384	17.99	20.93	16.74	18.00
1931	24,800	8,837	15,963	18.02	21.68	16.47	18.00
1932	20,000	7,179	12,821	18.04	20.36	16.95	18.00
1933	29,500	9,878	19,622	18.02	18.90	17.60	18.00
1934	28,900	9,365	19,535	17.91	18.47	17.65	18.00
1935	29,300	7,852	21,448	17.92	19.51	17.43	18.00
1936	35,400	10,147	25,253	17.98	18.54	17.19	18.00
1937	44,300	12,155	32,145	17.96	18.00	17.95	18.00
1938	27,300	10,379	16,921	16.76	17.92	16.12	17.51
1939	35,500	10,772	24,728	15.89	17.16	15.40	16.00
1940	40,900	13,042	27,858	15.98	17.47	15.37	16.00
1941	54,400	12,520	41,880	15.99	17.16	15.67	16.00
1942	50,100	10,943	39,157	16.01	19.26	15.29	16.00
1943	47,300	12,522	34,778	16.01	19.05	15.14	16.00
1944	56,300	12,236	44,064	16.00	18.72	15.38	16.00
1945	61,300	16,643	44,657	15.99	18.12	15.32	16.00
1946	66,100	21,590	44,510	16.01	18.16	15.14	16.00
1947	85,200	25,388	59,812	17.65	19.54	16.95	16.51
1948	89,600	26,779	62,821	18.00	21.20	16.91	18.00
1949	86,200	30,490	55,710	18.00	21.31	16.59	18.00
1950	104,000	30,951	73,049	18.99	21.48	17.98	19.02
1951	107,300	31,760	75,540	21.51	24.66	20.41	22.01
1952	110,925	33,515	77,410	21.57	25.70	20.17	22.01
1953	141,054	34,554	106,500	27.00	27.83	26.74	24.74
1954	142,014	50,362	91,652	26.65	30.62	24.88	26.50
1955	163,156	48,708	114,448	27.94	30.42	27.00	26.50
1956	150,356	48,305	102,051	26.49	29.25	25.36	26.50
1957	122,915	43,940	78,975	24.41	27.84	22.85	25.51
1958	109,272	39,975	69,765	23.53	25.04	22.75	23.50
1959	121,777	39,975	81,802	23.32	24.80	22.66	23.50

Table A-3. (Continued)

Year	Total value			Average value			Average posted price (7)
	All shipments (1)	Export shipments (2)	Domestic shipments (3)	All shipments (4)	Export shipments (5)	Domestic shipments (6)	
	(.. ..... *thousand dollars* ....)			(............... *dollars/long ton* ....)			
1960	115,494	40,880	74,614	23.09	23.02	23.12	23.50
1961	117,884	35,370	82,514	23.19	22.31	23.60	23.50
1962	107,069	35,496	71,573	21.77	23.09	21.18	23.50
1963	99,014	39,651	81,126	19.82	20.91	19.31	23.50
1964	120,777	39,651	81,126	20.01	20.64	19.71	24.50
1965	164,654	64,278	100,376	22.71	24.49	21.69	25.50
1966	201,292	78,759	122,533	26.07	33.86	22.71	26.75
1967	251,670	81,492	170,175	32.76	39.91	30.18	33.50

SOURCES: 1900 to 1932 data taken from U.S. Department of the Interior, Bureau of Mines (Geological Survey before 1924), *Mineral Resources of the United States* (Washington, D.C.: 1900–1932). Data after 1932 taken from U.S. Department of the Interior, Bureau of Mines, *Minerals Yearbook* (Washington, D.C.: 1932–33 to the present). Data on posted price taken from U.S. Department of Labor, Bureau of Labor Statistics, *Wholesale Prices*, Bulletin No. 320 for 1890–1912, Bulletin No. 493 for 1913–1928, annual bulletins for 1929–1948, summaries for 1947–1950 and 1951–1953, and annual bulletins for 1954–1967.

NOTES: Column (2) subtracted from Column (1) to obtain Column (3). Columns (4), (5), and (6) obtained by dividing total tons shipped, total tons exported, and total tons shipped minus total tons exported into Columns (1), (2), and (3) respectively. (Data on shipments taken from sources listed above, but not given in the table.)

APPENDIX B

Financial Data for Frasch Sulphur Producers

The investment and rate of return series for the period from 1919 to 1946 were taken from the United States Federal Trade Commission, *Report on the Sulphur Industry and International Cartels* (Washington, D.C., 1947). The Commission computed investment in the sulphur industry for each of the four producers. Annual balance sheet investment for each company at the end of the year was adjusted to more nearly reflect actual investment in the sulphur industry by eliminating items such as investment in governmental and other securities and surpluses arising from reappraisal of assets. The balance sheet information of Freeport Sulphur Company was adjusted to exclude its manganese and nickeliferous iron ore mining activities in Cuba. Similarly, net profits before taxes were adjusted for other income, other expenses, and like revenue items to obtain only the net revenue arising from sulphur operations. Average investment, computed from investment at the beginning and end of each year, was divided into net profits to obtain the rate of return for each company. Both before-tax and after-tax rates of return are shown for Texas Gulf Sulphur Company, Freeport Sulphur Company, and Jefferson Lake Sulphur Company. Since Duval did not publish before-tax profits, only after-tax rates of return are shown.

The investment and rate of return series for the period since 1946 were computed from information reported to the Securities and Exchange Commission and published in *Moody's Manual of Industrial Investments*. The average invested capital data represent the sum of the par value of capital stock, paid-in capital, retained earnings, and any surplus or reserve appropriated from retained earnings. No attempt was made to adjust for investment in activities other than sulphur mining or for income and expense arising from other than sulphur mining.

The compilation of an investment and profit series for the Frasch segment of the domestic sulphur industry is made easier by the fact that prior to 1950 only Freeport was engaged to any extent in activities other than the production of sulphur. During the 1950s, however, the four domestic Frasch producers diversified into other lines of activity. Nevertheless, sulphur mining remained the dominant activity of both Texas Gulf and Freeport through 1966 and of Jefferson Lake through 1962. (In 1964, Jefferson Lake was acquired by Occidental Petroleum Corporation.) However, over the past two decades sulphur mining has become an increasingly smaller element of Duval's operations. Presently, sulphur accounts for about 9 per cent of Duval's net income.

158

Both of the Mexican Frasch producers were engaged solely in the production of sulphur in the period prior to 1967. For Pan American, investment and profits are shown over the life of the firm from 1955 to 1966. For Gulf Sulphur, the data cover only the period from 1961 to 1966. While this firm began sulphur operations in 1956, it was not until 1961 that the firm showed a profit.

Table B-1. Summary of Investment, Profits, and Rates of Return on Investment for Texas Gulf Sulphur Company, 1919 to 1966

		Profit		Rate of return	
Year	Average invested capital*	Before federal taxes	After federal taxes	Before federal taxes	After federal taxes
	(. . . . . *thousand dollars* . . . .)			(. . . . *per cent* . . . . .)	
1919	5,159	994	968	19.27	18.76
1920	8,981	3,519	3,327	39.18	37.04
1921	11,261	1,941	1,866	17.24	16.58
1922	12,193	3,999	3,809	32.80	31.24
1923	12,867	4,972	4,683	38.64	36.39
1924	13,224	5,089	4,762	38.48	36.01
1925	13,260	6,028	5,627	45.46	42.44
1926	14,790	10,036	9,296	67.86	62.85
1927	17,687	13,110	12,002	74.12	67.86
1928	21,673	15,661	14,411	72.26	66.49
1929	27,589	17,624	16,185	63.88	58.66
1930	33,272	15,101	13,972	45.39	41.99
1931	35,791	9,772	8,943	27.30	24.99
1932	36,179	6,374	5,910	17.62	16.34
1933	37,506	7,957	7,444	21.21	19.85
1934	49,238	7,337	6,743	14.90	13.69
1935	59,739	8,178	7,468	13.69	12.50
1936	59,133	10,843	9,633	18.34	16.66
1937	59,131	12,864	11,589	21.76	19.60
1938	59,353	7,634	6,964	12.86	11.75
1939	59,349	8,922	7,847	15.03	13.22
1940	59,375	10,851	9,141	18.28	15.40
1941	58,770	12,966	9,016	21.05†	15.34
1942	58,642	13,449	8,779	21.36†	14.97
1943	58,789	14,416	7,966	22.40†	13.55
1944	58,721	16,820	9,620	25.66†	16.38
1945	59,089	18,193	9,993	27.24†	16.91
1946	61,485	21,590	15,240	31.40†	24.79
1947	65,940	30,665	21,665	46.50	32.86
1948	57,545	34,081	24,231	59.22	42.11
1949	50,273	33,238	23,863	66.12	47.47

Table B-1. (Continued)

Year	Average invested capital*	Profit Before federal taxes	Profit After federal taxes	Rate of return Before federal taxes	Rate of return After federal taxes
		Profit		Rate of return	
	Average invested capital*	Before federal taxes	After federal taxes	Before federal taxes	After federal taxes
	(........thousand dollars........)			(....per cent....)	
1950	57,367	38,889	25,889	67.79	45.13
1951	64,502	40,942	25,442	63.47	39.44
1952	68,830	38,612	25,112	56.10	36.48
1953	73,530	40,284	24,534	54.79	33.37
1954	83,576	46,245	30,555	55.33	36.56
1955	95,878	49,856	32,356	52.00	33.75
1956	106,139	42,386	28,136	39.93	26.51
1957	110,259	20,257	17,557	18.37	15.92
1958	111,987	16,883	13,383	15.08	11.95
1959	115,211	17,438	13,338	15.14	11.58
1960	119,304	17,434	12,684	14.61	10.63
1961	124,680	17,333	12,583	13.90	10.09
1962	132,352	15,887	12,137	12.00	9.17
1963	142,245	10,904	9,354	7.67	6.58
1964	151,448	15,060	11,556	9.94	7.63
1965	164,813	24,158	18,161	14.66	11.02
1966	199,164	39,928	28,096	20.05	14.11

SOURCES: U.S. Federal Trade Commission, *Report on the Sulphur Industry and International Cartels* (Washington: U.S. Government Printing Office, 1947) for years through 1946; and *Moody's Manual of Industrial Investments* (New York: Moody's Investors' Service, various issues) for 1947 to 1966.

* Beginning and end of year.

† Because of the high income and excess profits taxes during the war years, rates of return were computed by relating profit before federal taxes to the average invested capital adjusted for such tax provisions.

Table B-2. Summary of Investment, Profits, and Rates of Return on Investment for Freeport Sulphur Company, 1919 to 1966

Year	Average invested capital*	Profit		Rate of return	
		Before federal taxes	After federal taxes	Before federal taxes	After federal taxes
	(.........thousand dollars.........)			(....per cent.....)	
1919	9,632	1,185	1,105	12.30	11.48
1920	8,406	584	584	6.94	6.94
1921	8,446	(492)	(492)	(5.82)	(5.82)
1922	10,005	(253)	(253)	(2.53)	(2.53)
1923	12,175	770	770	6.33	6.33
1924	12,228	(326)	(326)	(2.65)	(2.65)
1925	11,772	750	750	6.37	6.37
1926	12,566	1,809	1,809	14.40	14.40
1927	13,830	3,925	3,736	28.38	27.01
1928	13,318	3,460	3,276	25.98	24.59
1929	11,621	4,216†	3,843†	36.28†	33.07†
1930	10,434	2,834	2,502	27.12	23.94
1931	10,364	2,177	1,918	21.00	18.51
1932	9,392	2,253	2,005	23.99	21.35
1933	10,807	2,714	2,479	25.11	22.94
1934	12,628	1,625	1,477	12.87	11.70
1935	12,363	1,642	1,492	13.28	12.07
1936	13,158	2,488	2,201	18.91	16.73
1937	14,122	2,713	2,443	19.21	17.30
1938	13,913	1,678	1,513	12.06	10.87
1939	14,927	1,828	1,622	12.25	10.87
1940	17,534	2,547	2,167	14.52	12.36
1941	18,933	3,281	2,581	16.85‡	13.63
1942	19,655	2,808	2,083	13.79‡	10.60
1943	19,340	3,310	2,425	16.43‡	12.54
1944	19,087	3,658	2,208	18.06‡	11.57
1945	20,156	4,902	2,902	22.40‡	14.40
1946	21,697	4,453	3,368	19.16‡	15.52
1947	29,924	3,446	2,723	11.52	9.10
1948	32,025	4,895	4,088	15.28	12.77
1949	34,583	7,190	5,882	20.79	17.01
1950	37,114	9,363	7,169	25.23	19.32
1951	39,453	8,728	6,415	22.12	16.26
1952	41,670	9,162	7,426	21.99	17.82
1953	44,801	10,786	8,706	24.08	19.43
1954	48,711	13,406	10,207	27.52	20.95

Table B-2. (Continued)

Year	Average invested capital*	Before federal taxes	After federal taxes	Before federal taxes	After federal taxes
		Profit		Rate of return	
		(........*thousand dollars*........)		(....*per cent*....)	
1955	57,737	15,163	12,599	26.26	21.82
1956	67,665	15,908	13,493	23.51	19.94
1957	73,387	16,928	13,124	23.07	17.88
1958	112,541	16,398	14,617	14.57	12.99
1959	151,672	16,097	15,332	10.61	10.11
1960	147,517	14,349	13,965	9.73	9.47
1961	142,902	13,374	13,135	9.36	9.19
1962	146,989	12,997	12,868	8.84	8.75
1963	152,866	13,596	12,894	8.89	8.43
1964	161,424	17,011	15,411	10.54	9.55
1965	172,512	26,047	21,923	15.10	12.71
1966	190,705	39,163	32,748	20.54	17.17

SOURCES: See table B-1.

* Beginning and end of year.

† 13 months.

‡ Because of the high income and excess profits taxes during the war years, rates of return were computed by relating the profit before federal income taxes to the average invested capital adjusted for such tax provisions.

( ) Loss.

Table B-3. Summary of Investment, Profits, and Rates of Return on Investment for Duval Corporation, 1930 to 1966

Year	Average invested capital*	Profit after federal taxes	Rate of return after federal taxes
	(..... *thousand dollars* .....)		*per cent*
1930	1,121	232	20.66
1931	1,341	198	14.76
1932	1,517	151	9.98
1933	1,307	(85)	(6.54)
1934	1,080	39	3.61
1935	1,196	78	6.54
1936	1,275	306	24.03
1937	1,571	221	14.07
1938	1,923	379	19.69
1939	2,421	652	26.92
1940	2,817	581	20.62
1941	2,820	709	25.13
1942	2,743	795	28.99
1943	2,616	809	30.91
1944	2,566	799	31.12
1945	2,619	786	30.00
1946	2,832	1,012	35.7:
1947	4,183	1,073	25.65
1948	3,928	764	19.45
1949	3,609	781	21.64
1950	5,510	1,007	18.28
1951	8,448	1,187	14.05
1952	10,474	2,681	25.60
1953	12,289	3,045	24.78
1954	14,182	3,072	21.66
1955	15,996	3,063	19.15
1956	17,531	2,507	14.30
1957	21,339	3,110	14.57
1958	26,501	2,645	9.98
1959	28,845	2,349	8.14
1960	30,691	4,595	14.97
1961	33,415	4,104	12.28
1962	35,863	4,463	12.44
1963	39,756	4,364	10.98
1964	45,240	5,442	12.03
1965	58,071	9,604	16.54
1966	73,919	11,036	14.93

SOURCES: See table B-1.
* Beginning and end of year.
( ) Loss.

Table B-4. *Summary of Investment, Profits, and Rates of Return on Investment for Jefferson Lake Sulphur Company, 1933 to 1962*

Year	Average invested capital*	Profit Before federal taxes	Profit After federal taxes	Rate of return Before federal taxes	Rate of return After federal taxes
	·(........thousand dollars........)			(....per cent.....)	
1933	2,351	1,019	841	43.33	35.77
1934	2,554	1,005	864	39.36	33.85
1935	2,389	.....	132	.....	5.51
1936	n.a.	n.a.	n.a.	n.a.	n.a.
1937	2,188	(11)	(11)	(.51)	(.51)
1938	2,469	977	821	39.56	33.27
1939	2,819	1,529	1,325	54.23	47.01
1940	2,830	1,075	812	37.98	28.70
1941	2,676	323	280	12.09	10.46
1942	2,374	(328)	(328)	(13.81)	(13.81)
1943	2,142	149	138	6.78	6.46
1944	2,207	431	358	19.54	16.20
1945	2,682	424	359	15.80	13.40
1946	3,191	395	368	12.22	11.52
1947	3,349	431	382	12.87	11.41
1948	3,637	622	420	17.10	11.55
1949	3,858	1,027	525	26.62	13.61
1950	4,349	1,898	1,246	43.64	28.65
1951	5,056	2,150	1,321	42.52	26.13
1952	5,684	2,086	1,232	36.70	21.67
1953	6,307	2,288	1,487	36.28	23.58
1954	7,228	3,841	2,184	53.14	30.22
1955	8,136	2,652	1,711	36.60	21.03
1956	8,683	2,569	1,608	29.59	18.52
1957	9,391	1,350	1,243	14.38	13.24
1958	10,892	(538)	(659)	(4.94)	(6.05)
1959	11,316	978	(1,329)	8.64	(11.74)
1960	10,789	867	181	8.04	1.68
1961	9,916	2,104	1,006	21.22	10.15
1962	9,538	2,060	1,026	21.60	10.76

SOURCES: See table B-1.
n.a.—Data not available.
* Beginning and end of year.
( ) Loss.

Table B-5. *Summary of Investment, Profits, and Rates of Return on Investment for Pan American Sulphur Company, 1955 to 1966*

Year	Average invested capital*	Profit Before federal taxes	Profit After federal taxes	Rate of return Before federal taxes	Rate of return After federal taxes
	(........thousand dollars........)			(....per cent.....)	
1955	5,725	623	378	10.88	6.60
1956	7,236	3,775	2,122	52.17	29.33
1957	10,135	6,319	3,455	62.35	34.09
1958	13,188	6,878	3,575	52.15	27.11
1959	16,512	6,547	3,462	39.65	20.97
1960	19,155	6,149	3,119	32.10	16.28
1961	19,628	5,465	2,415	27.84	12.30
1962	20,067	6,846	3,082	34.12	15.36
1963	21,037	8,029	3,467	38.17	16.48
1964	23,254	11,200	5,120	48.16	22.02
1965	25,432	8,230	4,330	32.36	17.03
1966	28,780	16,389	8,303	56.95	28.85

SOURCE: Pan American Sulphur Company, Annual Reports.
* Beginning and end of year.

Table B-6. *Summary of Investment, Profits, and Rates of Return on Investment for Gulf Sulphur Corporation, 1961 to 1966*

Year	Average invested capital*	Profit Before federal taxes	Profit After federal taxes	Rate of return Before federal taxes	Rate of return After federal taxes
	(....... thousand dollars ....)			(....per cent.....)	
1961	5,145	254	254†	4.94	4.94†
1962	5,006	1,614	1,139	32.24	22.75
1963	6,082	1,143	916	18.69	15.06
1964	6,978	1,014	767	14.53	10.99
1965	8,307	2,529	1,716	30.44	20.66
1966	10,476	3,200	2,374	30.55	22.66

SOURCE: Gulf Sulphur Corporation, Annual Reports.
* Beginning and end of year.
† No income tax paid in 1961.

Bibliography

I. Books, Monographs, and Theses

Acevedo Escobedo, Antonio. *El azufre en Mexico.* Mexico, D.F.: Editorial Cultura, 1956.

Adams, Walter, and Gray, Horace M. *Monopoly in America: The Government as Promoter.* New York: The Macmillan Company, 1955.

Arthur D. Little, Inc. *The Free World Sulphur Outlook.* 1966.

Bain, Joe S. *The Economics of the Pacific Coast Petroleum Industry.* Berkeley: University of California Press, 3 vols., 1944, 1945, 1947.

————. *Barriers to New Competition.* Cambridge: Harvard University Press, 1956.

Barger, Harold, and Schurr, Sam H. *The Mining Industries, 1899–1939.* New York: National Bureau of Economic Research, 1944.

Barnett, Harold J., and Morse, Chandler. *Scarcity and Growth.* Baltimore: The Johns Hopkins Press for Resources for the Future, Inc., 1965.

Baruch, Bernard M. *Baruch, My Own Story.* New York: Henry Holt and Company, 1957.

Bernstein, Marvin D. *The Mexican Mining Industry.* New York: State University of New York, 1965.

Bixby, David W.; Tisdale, Samuel L.; and Rucker, Delbert R. *Adding Plant Nutrient Sulphur to Fertilizer.* Washington, D.C.: The Sulphur Institute, Technical Bulletin no. 10, 1964.

Brese, W. G. *An Analysis of the Sulphur Industry in Alberta.* Calgary, Alberta: Research Council of Alberta, Information Series, no. 38, 1962.

Chamberlin, Edward H. *The Theory of Monopolistic Competition.* 8th ed. Cambridge: Harvard University Press, 1962.

Clow, Archibald, and Clow, Nan L. *The Chemical Revolution.* London: The Batchworth Press, 1952.

Duecker, Werner W., and West, James R. *The Manufacture of Sulfuric Acid.* New York: Reinhold Publishing Corporation, 1959.

Earp, M. H. "The Frasch Sulphur Industry of Mexico." Master's thesis, Southern Methodist University, 1960.

Edwards, Edgar O., and Bell, Philip W. *The Theory and Measurement of Business Income.* Berkeley: University of California Press, 1961.

Fairlie, Andrew M. *Sulfuric Acid Manufacture.* New York: Reinhold Publishing Corporation, 1936.

Faith, W. L.; Keyes, Donald B.; and Clark, Ronald L. *Industrial Chemicals.* 2nd ed. New York: John Wiley and Sons, Inc., 1957.

Fellner, William. *Competition Among the Few.* New York: Alfred A. Knopf, 1949.

167

Hamilton, Daniel C. *Competition in Oil.* Cambridge: Harvard University Press, 1958.

Hansen, Eric J. *Dynamic Decade.* Toronto, Canada: McClelland and Stewart, Ltd., 1958.

Haynes, Williams. *The Stone That Burns.* New York: D. Van Nostrand Company, Inc., 1942.

———. *Brimstone: The Stone That Burns.* Princeton: D. Van Nostrand Company, Inc., 1959.

Kaysen, Carl. *United States v. United Shoe Machinery Corporation.* Cambridge: Harvard University Press, 1956.

Kreps, Theodore J. *The Economics of the Sulfuric Acid Industry.* Stanford: Stanford University Press, 1938.

Landsberg, Hans H.; Fischman, Leonard L.; and Fisher, Joseph L. *Resources in America's Future.* Baltimore: The Johns Hopkins Press for Resources for the Future, Inc., 1963.

Lehmann, Glenn Albert. "The Market for Sulphur: A Study in Duopoly." Ph.D. dissertation, Harvard University, 1953.

Liebhafsky, H. H. *The Nature of Price Theory.* Homewood, Illinois: The Dorsey Press, Inc., 1963. Rev. ed., 1968.

Liebig, Justus. *Familiar Letters on Chemistry.* Edited by John Gardner. New York: J. Winchester, New World Press, 1843.

Markham, Jesse W. *Competition in the Rayon Industry.* Cambridge: Harvard University Press, 1952.

———. *The Fertilizer Industry.* Nashville: The Vanderbilt University Press, 1958.

Mason, Edward S. *Economic Concentration and the Monopoly Problem.* Cambridge: Harvard University Press, 1957.

McKie, James W. *Tin Cans and Tin Plate.* Cambridge: Harvard University Press, 1959.

McLean, John G., and Haigh, Robert William. *The Growth of Integrated Oil Companies.* Boston: Division of Research, Graduate School of Business Administration, Harvard University, 1954.

Montgomery, R. H. *The Brimstone Game.* Manchaca, Texas: The Chaparral Press, 1949. First published, 1940.

Morrison, Thurmond L. "The Economics of the Sulphur Industry." Ph.D. dissertation, University of Texas, 1939.

Peck, Merton J. *Competition in the Aluminum Industry, 1945–1958.* Cambridge: Harvard University Press, 1961.

Phillips, Charles F., Jr. *Competition in the Synthetic Rubber Industry.* Chapel Hill, North Carolina: The University of North Carolina Press, 1961.

Robinson, Joan. *The Economics of Imperfect Competition.* London: Macmillan and Company, Ltd., 1961.

Schumpeter, Joseph A. *Capitalism, Socialism, and Democracy.* 3rd ed. New York: Harper and Brothers Publishers, 1950.

Scott, Anthony. *Natural Resources: The Economics of Conservation.* Toronto, Canada: The University of Toronto Press, 1955.

Shubik, Martin. *Strategy and Market Structure.* New York: John Wiley and Sons, Inc., 1959.

Stigler, George J. *The Theory of Price.* New York: The Macmillan Company, 1946. Rev. ed., 1952.

———. *Capital and Rates of Return in Manufacturing Industries.* Princeton: Princeton University Press for the National Bureau of Economic Research, 1963.

Stocking, George W., and Watkins, Myron W. *Cartels or Competition?* New York: The Twentieth Century Fund, 1948.

Stoddard, X. T. "Scout Memo, High Island Dome Sulphur Mine, Galveston County, Texas." Memorandum, Humble Oil and Refining Company, 1959.

Swagger, William L. "The Paley Report in Review: Sulphur." Unpublished monograph, Battelle Memorial Institute, 1961.

Texas Gulf Sulphur Company. *Facts About Sulphur.* New York: Texas Gulf Sulphur Company, 1958.

————. *Modern Sulphur Mining.* New York: Texas Gulf Sulphur Company, 1961.

Truett, Dale B. "Sulphur and the Development of a Chemical Fertilizer Industry in Mexico." Ph.D. dissertation, University of Texas, 1967.

Tuller, William N., ed. *The Sulphur Data Book.* New York: McGraw-Hill Book Company, Inc., 1954.

Viner, Jacob. *Dumping: A Problem in International Trade.* Chicago: University of Chicago Press, 1923.

II. Public Documents

Federal Trade Commission. *Report on the Sulphur Industry and International Cartels.* 1947.

President's Materials Policy Commission. *Resources for Freedom.* 1952.

United States Congress. *Hearings before the Temporary National Economic Committee.* 76th Congress, 1st Session, Part V, 1939.

United States Department of the Interior, Bureau of Mines. *Minerals Yearbook.* Annual issues.

————. *Mineral Trade Notes.* Monthly issues.

United States House of Representatives. *President's 1963 Tax Message.* 88th Congress, 1st Session, Part I, 1963.

United States Tariff Commission. *Information Concerning the Pyrites and Sulphur Industry.* 1919.

————. "Industrial Readjustments of Certain Mineral Industries Affected By the War." *Tariff Information Series, Number 21.* 1920.

III. Articles and Periodicals

Ambrose, Paul M. "Sulphur and Pyrites." In *Mineral Facts and Problems, 1965 Edition.* United States Department of the Interior, Bureau of Mines, Bulletin 630, 1965.

Adelman, M. A. "Effective Competition and the Antitrust Laws." *Harvard Law Review* (1948), 1289–1350.

Bain, Joe S. "Workable Competition in Oligopoly: Theoretical Considerations and Some Empirical Evidence." *American Economic Review* (1950), 35–47.

Blair, John M. "Means, Thorp, and Neal on Price Inflexibility." *Review of Economics and Statistics* (1956), 427–35.

Brooks, David B. "The Supply of Individually Mined Minor Metals and Its Implications for Subsidy Programs." *Land Economics* (1964), 18–24.

Butterworth, C. D., and Schwab, J. W. "Sulfur Mining as a Processing Industry." *Industrial and Engineering Chemistry* (1938), 746–51.

Carsey, J. Ben. "Geology of Gulf Coastal Area and Continental Shelf." *Bulletin of the American Association of Petroleum Geologists* (March 1950), 361–85.

Clark, J. M. "Towards a Concept of Workable Competition." *American Economic Review* (1940), 241–56.

DeGolyer, E. "Origin of North American Salt Domes." *Bulletin of the American Association of Petroleum Geologists* (August 1925), 831–74.

Deschamps, Federico. "Los domas del sal del Istmo de Tehuantepec desde el punto de vista de su importancia económica." *Revista Mexicana de Ingenieria y Arquitectura* (November and December 1937), 711–23 and 777–99.

Doak, John. "Liquid-Sulphur Distribution." *The Oil and Gas Journal* (June 24, 1963), 101.

Estep, James W.; McBride, Guy T.; and West, James R. "The Recovery of Sulphur from Sour Natural and Refinery Gases." In *Advances in Petroleum Chemistry and Refining*, edited by John J. McKetta, Jr. New York: Interscience Publishers, 1962.

Fellner, William. "Collusion and Its Limits Under Oligopoly." *American Economic Review* (1950), 54–62.

Gittinger, L. B., Jr. "Sulphur." *The Engineering and Mining Journal* (February 1964), 151.

———. "Sulphur." *The Engineering and Mining Journal* (February 1967), 170.

———. "Sulphur." *The Engineering and Mining Journal* (March 1969), 160C–160F.

Graff, R. A. "Elemental Sulphur from Petroleum Gases." *The Oil and Gas Journal* (17 February 1949), 103–4.

Gray, L. C. "Rent Under the Assumptions of Exhaustibility." *Quarterly Journal of Economics* (May 1914), 446–89.

Hanna, Marcus A., and Wolf, Albert G. "Texas and Louisiana Salt Dome Cap Rock Minerals." In *Gulf Coast Oil Fields*, edited by Donald C. Barton and George Sawtelle. Tulsa, Oklahoma: The American Association of Petroleum Geologists, 1936.

Hartley, Burton. "The Petroleum Geology of the Isthmus of Tehuantepec." *Economic Geology* (October–November 1917), 581–88.

Hawkins, M. E., and Jink, C. F. "Salt Domes in Texas, Louisiana, Mississippi, Alabama, and Offshore Tidelands: A Survey." United States Department of the Interior, Bureau of Mines, *Information Circular 8313* (1966).

Herfindahl, Orris C. "Some Fundamentals of Mineral Economics." *Land Economics* (1955), 131–38.

Hotelling, Harold. "The Economics of Exhaustible Resources." *Journal of Political Economy* (1931), 137–75.

Hunt, Walter F. "The Origin of the Sulphur Deposits of Sicily." *Economic Geology* (1915), 543–79.

Huntley, Stirling. "Oil Development on the Isthmus of Tehuantepec." *Transactions of the American Institute of Mining and Metallurgical Engineers* (1923), 1150–66.

Jones, H. H., and Graff, R. A. "West Texas' First Recovery Unit." *The Oil and Gas Journal* (April 21, 1952), 122–23 and 153–54.

Kearney, John H. "A New Empire of Frasch Process Sulphur Is Rising from the Jungles of Mexico." *Engineering and Mining Journal* (January 1955), 72–77.

Kelly, F. J. "Sulphur Production and Consumption in Eight Western States."

United States Department of the Interior, Bureau of Mines, *Information Circular 8094* (1962).

Lee, C. O.; Bartlett, Z. W.; and Feierabend, R. H. "The Grand Isle Mine." *Mining Engineering* (June 1960), 578–90.

Lundy, W. T. "Sulphur and Pyrites." In *Industrial Minerals and Rocks*. 2nd ed. New York: American Institute of Mining and Metallurgical Engineers, 1949.

————. "Known and Potential Sulphur Resources of the World." *Industrial and Engineering Chemistry* (November 1950), 2200.

Mason, Donald B. "The Sulphur Industry." *Industrial and Engineering Chemistry* (1938), 740–46.

Mason, Edward S. "The Political Economy of Resource Use." In *Perspectives on Conservation*, edited by Henry Jarrett. Baltimore: The Johns Hopkins Press for Resources for the Future, Inc., 1958.

Means, Gardiner C. "Basic Characteristics of the American Economy." In National Resources Committee, *The Structure of the American Economy*. Part I (1939).

Nelson, James R. "Percentage Depletion and National Security." In *Federal Tax Policy for Economic Growth and Stability*. Joint Committee Print of the 84th Congress, 1st Session, 1955.

Netzeband, F. F.; Early, Thomas R.; Ryan, J. P.; and Miller, W. C. "Sulphur Resources and Production in Texas, Louisiana, Missouri, Oklahoma, Arkansas, Kansas, and Mississippi, and Markets for the Sulphur." United States Department of the Interior, Bureau of Mines, *Information Circular 8222* (1964).

O'Hanlan, Thomas. "The Great Sulphur Rush." *Fortune* (March 1968), 109.

Price, K. T. "Freeport Mines Sulphur by Boat and Barge at Bay Ste. Elaine." *Engineering and Mining Journal* (December 1952), 98–102.

Reed, Robert M., and Updegraff, Norman C. "Removal of Hydrogen Sulphide from Industrial Gases." *Industrial and Engineering Chemistry* (1950), 2269–77.

Rogers, Sherburne. "The Intrusive Origin of the Gulf Coast Salt Domes." *Economic Geology* (September 1918), 447–85.

Sawtelle, George. "Salt Dome Statistics." In *Gulf Coast Oil Fields*, edited by Donald C. Barton and George Sawtelle. Tulsa, Oklahoma: The American Association of Petroleum Geologists, 1936, 109–18.

Shearon, Will H., Jr., and Pollard, J. H. "Modern Sulphur Mining." In *Modern Chemical Processes*, edited by William J. Murphy. vol. 2. New York: Reinhold Publishing Corporation, 1952, 219–29.

Sheehan, Robert. "The 'Little Mothers' and Pan American Sulphur." *Fortune* (July 1960), 96–103.

Shibler, B. K., and Hovey, M. S. "Processes for Recovering Sulphur from Secondary Source Materials." United States Department of the Interior, Bureau of Mines, *Information Circular 8076* (1962).

Sosnick, Stephan H. "A Critique of Concepts of Workable Competition." *Quarterly Journal of Economics* (1958), 380–423.

Villard, H. H. "Competition, Oligopoly, and Research." *Journal of Political Economy* (1958), 483–97.

Villarillo, Juan D. "Algunas Regiones Petrolíferas de Mexico." *Boletin del Instituto Geologico de Mexico* (1908), 69–92.

Viner, Jacob. "Cost Curves and Supply Curves." *Zeitschrift fur National-Okonomie* (1932), 23–46.

Wells, A. E., and Fogg, D. E. "The Manufacture of Sulphuric Acid in the United States." United States Department of the Interior, Bureau of Mines, *Bulletin 184* (1920).

Williamson, Oliver E. "Innovation and Market Structure." *Journal of Political Economy* (February 1965), 67–73.

Wionczek, Miguel. "Foreign-Owned Export-Oriented Enclave in a Rapidly Industrializing Economy: Sulphur Mining in Mexico." In *El nacionalismo mexicano y la inversión extranjera.* Mexico, D.F.: Siglo XXI Editores, 1967.

Wolf, Albert C. "Marketing of Sulphur." *Engineering and Mining Journal* (July 1, 1922), 19–22.

Zimmerman, J. B., and Thomas, Eugene. "Sulphur in West Texas: Its Geology and Economics." The University of Texas at Austin, Bureau of Economic Geology, *Geological Circular 69–2* (1969).

IV. Personal Interviews

————. Personal interview with William Amos, Chief, Division of Agricultural and Inorganic Chemicals, Business and Defense Services Administration, United States Department of Commerce, Washington, D.C., August 5, 1966.

————. Personal interview with Clarence O. Babcock, Commodity Specialist (Sulphur), Minerals Division, United States Department of the Interior, Bureau of Mines, Washington, D.C., August 1964.

————. Personal interview with David W. Bixby, Chemical Engineer, The Sulphur Institute, Washington, D.C., August 3–4, 1966.

————. Personal interview with Hal H. Bybee, Geologist, Continental Oil Company, Houston, Texas, February 1964.

————. Personal interview with J. C. Carrington, Vice President, Freeport Sulphur Company, New York City, August 15, 1966.

————. Personal interview with Shelby M. Darbishire, Executive Vice President, Gulf Sulphur Company, Houston, Texas, May 21, 1964.

————. Personal interviews with Raymond Devine, Vice President, Pan American Sulphur Company, Houston, Texas, July 24, 1963 and January 30, 1964.

————. Personal interview with Eugene Germain, Vice President, Duval Corporation, Houston, Texas, February 1964.

————. Personal interview with Edward Getzin, Chief, Industrial and Strategic Minerals Division, United States Department of State, Washington, D.C., August 11, 1966.

————. Personal interview with William C. Hawk, Geologist, Continental Oil Company, Houston, Texas, February 1964.

————. Personal interview with I. E. McKeever, Jr., General Manager of Mines, Texas Gulf Sulphur Company, Newgulf, Texas, March 1964.

————. Personal interview with Delbert L. Rucker, Director of Information, The Sulphur Institute, Washington, D.C., August 9, 1966.

————. Personal interview with H. W. Strickland, Senior Vice President, Texas Gulf Sulphur Company, Houston, Texas, February 14, 1964.

————. Personal interview with H. C. Webb, President, Pan American Sulphur Company, Houston, Texas, January 1964.